UNIVERSAL THEORY FOR STRONG LIMIT THEOREMS OF PROBABILITY

UNIVERSAL THEORY FOR STRONG LIMIT THEOREMS OF PROBABILITY

A.N.Frolov

St. Petersburg State University, Russia

NEW JERSEY • LONDON • SINGAPORE • BEIJING • SHANGHAI • HONG KONG • TAIPEI • CHENNAI • TOKYO

Published by

World Scientific Publishing Co. Pte. Ltd.

5 Toh Tuck Link, Singapore 596224

USA office: 27 Warren Street, Suite 401-402, Hackensack, NJ 07601

UK office: 57 Shelton Street, Covent Garden, London WC2H 9HE

Library of Congress Control Number: 2019044572

British Library Cataloguing-in-Publication Data
A catalogue record for this book is available from the British Library.

UNIVERSAL THEORY FOR STRONG LIMIT THEOREMS OF PROBABILITY

ISBN 978-981-121-282-6

For any available supplementary material, please visit
https://www.worldscientific.com/worldscibooks/10.1142/11625#t=suppl

Printed in Singapore

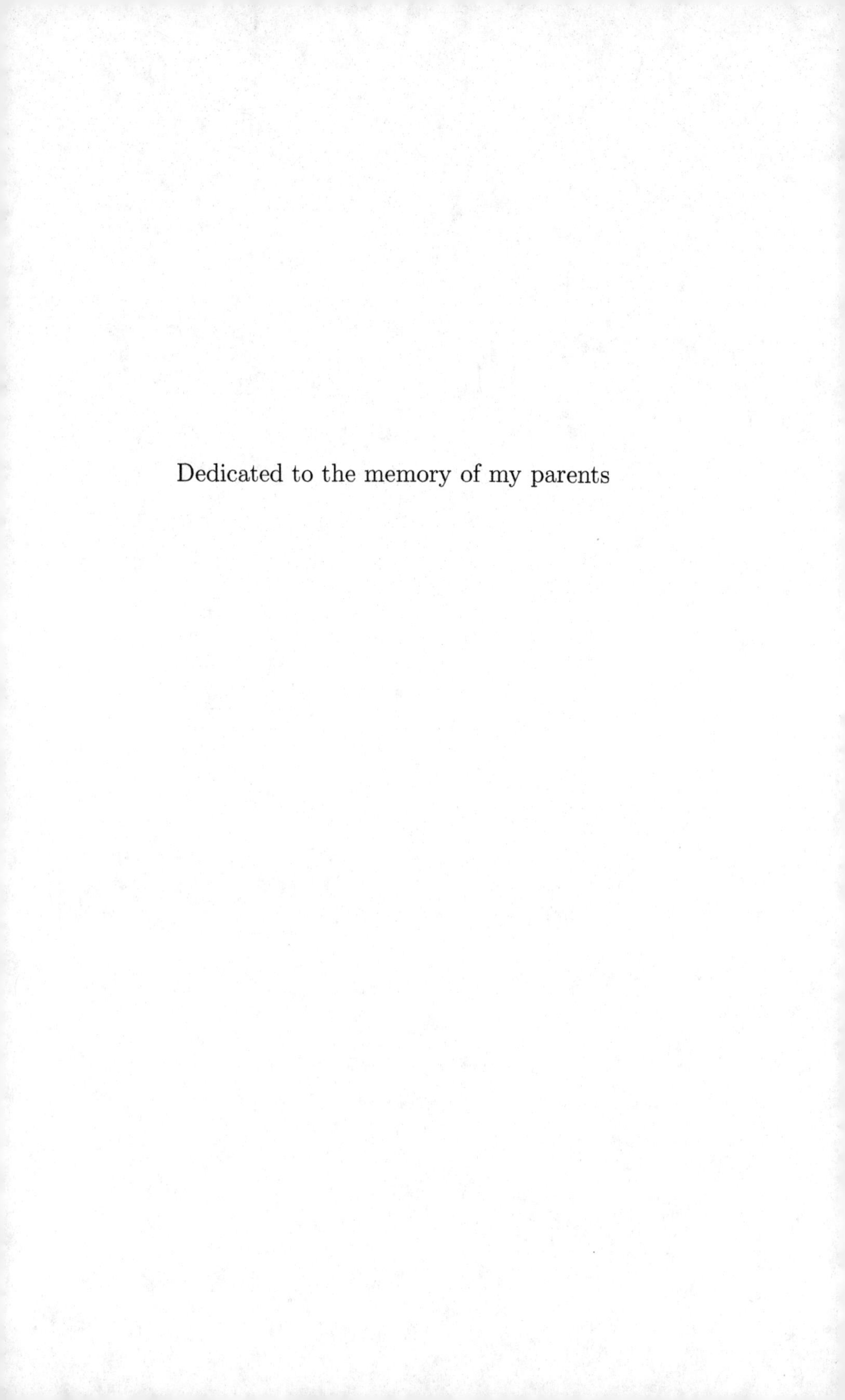

Dedicated to the memory of my parents

Preface

Limit theorems form an immanent part of probability and statistics. Strong limit theorems for sums of independent random variables are of essential interest in there. In particular, the strong law of large numbers provides the almost surely convergence of frequencies of events to their probabilities. Therefore, the abstract probability theory based on Kolmogorov's axiomatic approach describes laws of nature. In statistics, it yields desired properties of estimators and tests.

A significant attention of investigators was first paid to the strong law of large numbers and the law of the iterated logarithm. Later various results for increments of sums of independent random variables have been derived. The Kolmogorov strong law of large numbers, the Hartman–Wintner law of the iterated logarithm, the Erdős–Rényi and Shepp laws and the Csörgő–Révész laws are famous achievements for the case of independent, identically distributed random variables. Numerous generalizations and augments of these results have further been obtained in hundreds papers published on this subject for now.

In this book, we present a unified approach to strong limit theorems. It turns out that the results mentioned above are partial cases of general laws which we call universal strong laws. Besides sums of independent, identically distributed random variables, we derive such laws for processes with independent increments and renewal processes as well. We also concern with strong laws for some maxima from sums of independent random variables over head runs and monotone blocks. Note that an analysis of probabilities of large deviations is the best tool to prove strong limit theorems. Hence, the text includes corresponding methods and results of the large deviations theory which are of independent interest. Finally, mention that similar universal theories exist for independent, non-identically distributed random variables, random fields and some classes of stochastic processes.

The book is organized as follows. In Chapter 1, we start with a survey of results, methods and problems related to strong limit theorems. Further, we discuss relationships between universal strong laws and large deviations. Chapter 2 is devoted to the large deviations theory for sums of independent, identically distributed random variables. We discuss methods of conjugate distributions and truncations and derive logarithmic asymptotics for large deviations probabilities. We first consider random variables with a finite variation. Further, we turn to random variables from domains of attraction of the normal law and completely asymmetric stable laws with exponent from $(1,2)$. In Chapter 3, we describe a universal theory of strong limit theorems for sums of independent, identically distributed random variables. We start with universal strong laws. From them, we derive the Erdős–Rényi and Shepp laws, the Csörgő–Révész laws, the law of the iterated logarithm and the strong law of large numbers. We consider random variables with finite variations and those from domains of attraction mentioned above. Finally, we establish an optimality of one-sided moment assumptions which we deal with. In Chapters 4 and 5, we describe universal theories for homogeneous processes with independent increments and renewal processes, correspondingly. In Chapter 6, the universal theory is discussed for increments of sums of independent random variables over head runs and monotone blocks. In there, we deal with functionals those are of interest in game settings. We follow the pattern of Chapter 3 in Chapters 4–6. Staring with the universal laws, we derive all spectrum of strong limit theorems from them. In fact, we show that there are universal laws which manage the behaviours of increments for a variety of random sequences and processes.

I hope that this monograph will be useful for scientists, teachers and students being interested in probability theory.

I thank very much all my colleagues from Chair of Probability Theory and Mathematical Statistic of Saint–Petersburg State University and, especially, Prof. V.V.Petrov, Prof. A.I.Martikainen and Prof. V.B.Nevzorov for support and cooperation in many years.

St.Petersburg,
September, 2018

Andrei Frolov
Professor of Saint–Petersburg State University

Acronyms

i.i.d. independent, identically distributed

d.f. distribution function

c.f. characteristic function

m.g.f. moment generating function

a.s. almost surely

w.p. 1 with probability 1

i.o. infinitely often

LDT large deviations theory

SLLN strong law of large numbers

LIL law of the iterated logarithm

CLT central limit theorem

$\square$ end of a proof

$\mathbf{R}$ the set of real numbers

$\mathbf{N}$ the set of natural numbers

$a_n = o(b_n)$ means that $a_n/b_n \to 0$

$a_n = O(b_n)$ means that $\limsup |a_n|/b_n < \infty$

$a_n \sim b_n$ means that $a_n/b_n \to 1$

$I_B(x)$ the indicator of a Borel set B

I_B the indicator of an event B

DX the variation of a random variable X

$DN(\alpha)$ domain of normal attraction of the asymmetric stable law with exponent $\alpha > 1$

$D(\alpha)$ domain of non-normal attraction of the asymmetric stable law with exponent $\alpha > 1$

SV_a the set of slowly varying at a functions

RV_a the set of regularly varying at a functions

$f^{-1}(x)$ the inverse function to $f(x)$

$\#B$ the number of elements of a finite set B

$[x]$ the integer part of x

Contents

Chapter 1

Strong Laws and Large Deviations

Abstract. We start with a survey of strong laws and theory of large deviations in probability and statistics. Corresponding methods are discussed as well. Further, we give a universal approach to strong limit theorems that includes the SLLN, the LIL, the Erdős–Rényi law, the Shepp law and the Csörgő–Révész laws. We prove universal strong laws without a specification of norming sequences by an application of a large deviations method.

1.1 Strong Limit Theorems of Probability Theory: Results, Problems and Methods

In probability, one deals with mathematical models of random experiments (or observations) those are defined by two properties. First, all possible results of such experiment are known, but the result of a single one can not be predicted. Second, the empirical law of stability of frequencies is observed in long series of these experiments. For example, one can not predict the result of a single coin tossing, but one can see that the numbers of occurrences of heads and tails are relatively close after ten thousands of a symmetric coin tossing. So, frequencies of head (ratios of head numbers to numbers of experiments) converge to 0.5 as numbers of coin tossing tend to infinity. This is the law of stability of frequencies in our example.

The probability theory is a mathematical one and, consequently, it is based on axioms similarly to Euclidean geometry, for instance. Now Kolmogorov's axiomatic approach is usually applied. Hence, mathematical models of random experiments are measurable spaces with probability measures. To be sure that our models are relevant, we need a theorem corresponding to the empirical law of stability of frequencies. This is the strong law of large numbers (SLLN).

Let $X, X_1, X_2, \ldots$ be a sequence of independent, identically distributed (i.i.d.) random variables with a finite mean. Put $S_0 = 0$ and

$$S_n = X_1 + X_2 + \cdots + X_n.$$

By the Kolmogorov SLLN, we have

$$\frac{S_n}{n} \to EX \quad \text{a.s.} \tag{1.1}$$

Here and in the sequel, all limits are taken as $n \to \infty$ if it is not pointed otherwise and a.s. means almost surely. Note that $S_n/n \to c$ a.s. implies $E|X| < \infty$.

E.Borel has first proved SLLN for Bernoulli trails when $P(X = 1) = 1 - P(X = 0) = p \in (0, 1)$. If $p = 0.5$ and X_i is the number of heads in i-th repetition of a coin tossing, then (1.1) corresponds to the law of stability of frequencies in the above example. Moreover, since S_n/n is the frequency of head (or tail), sums S_n are basic objects of the theory. Applying the SLLN in the numbers theory, E.Borel has also proved the normality of almost all (with respect to the Lebesgue measure) real numbers (see [Feller (1971)] and [Lamperti (1996)] for details).

In statistics, one checks correspondences of probabilistic models and collections of data obtained from random experiments. For example, one could check hypothesis of symmetry of the coin using results of ten thousands of coin tossing. Strong laws imply properties of estimators and statistics which statistical tests are based on. The Borel SLLN yields the following Glivenko–Cantelli theorem:

$$\sup_{x \in \mathbf{R}} |F_n(x) - F(x)| \to 0 \quad \text{a.s.},$$

where $F_n(x)$ is the empirical distribution function (d.f.) for sample $X_1, \ldots, X_n$ and $F(x)$ is the d.f. of X. (See [Loève (1963)] for details.) Unbounded growths of chi square statistics under alternative hypotheses follow from the Borel SLLN and define the structure of the test. The Kolmogorov SLLN yields the latter for all tests based on statistics those are sums of i.i.d. random variables. Moreover, the Kolmogorov SLLN provides the consistency of empirical moments, for example, that of the empirical mean and variation. Hence, SLLN is a basis of statistics as well.

The next famous a.s. result is the law of the iterated logarithm (LIL). By the Hartman–Wintner LIL, if $EX = 0$ and $EX^2 = 1$, then

$$\limsup \frac{S_n}{\sqrt{2n \log\log n}} = 1 \quad \text{a.s.} \tag{1.2}$$

Note that one can replace S_n by $|S_n|$ in relation (1.2). Later, it was independently established in [Martikainen (1980)], [Rosalsky (1980)] and [Pruitt (1981)] that $EX = 0$ and $EX^2 = 1$ are also necessary conditions for (1.2).

For the Bernoulli case with $P(X = -1) = 1 - P(X = 1) = 0.5$, the SLLN yields $S_n = o(n)$ a.s. Further a.s. bounds for S_n have been obtained consequently by F.Haussdorff, G.Hardy and J.Littlewood and, finally, by A.Khinchine who has proved LIL. In particular, they also concerned with Rademaher functions those are i.i.d Bernoulli random variables on $[0, 1]$ with the Lebesgue measure. (See [Lamperti (1996)] for details.)

The Hartman–Wintner LIL yields the following property of trajectories of random walks $\{S_n\}$: with probability 1 (w.p. 1), for every $\varepsilon > 0$ the inequalities $|S_n| \le (1 + \varepsilon)\sqrt{2n \log\log n}$ hold for all sufficiently large n. The latter gives bounds for convergence rates of statistical estimators to parameters. For the sample mean $\overline{X} = S_n/n$, it follows that w.p. 1, for every $\varepsilon > 0$ we get $|\overline{X} - EX| \le (1 + \varepsilon)\sqrt{2 \log\log n/n}$ for all sufficiently large n.

Detailed discussions on the SLLN and LIL, a history and references may be found in [Petrov (1975, 1987, 1995)].

Relation (1.2) fails when $EX^2 = \infty$. Nevertheless, similar results may hold true for X from domain of attraction of a completely asymmetric stable law with exponent $\alpha \in (1, 2]$. For $\alpha = 2$, it is the Gaussian distribution. Remember that stable laws are limit distributions for normalized sums S_n with respect to weak convergence. The simplest case is the Lévy central limit theorems (CLT). LIL (1.2) is related with the asymptotic normality of $S_n/\sqrt{n}$. A similar situation holds for non-Gaussian case. Distributions of normalized S_n are close or coincide with a stable law. LIL for this case may be found in [Mijnheer (1974)] and references therein. For X from the domain of normal attraction of the completely asymmetric stable law with $\alpha \in (1, 2)$ and $EX = 0$, the LIL holds with the norming $cn^{1/\alpha}(\log\log n)^{(\alpha-1)/\alpha}$ which turns to that from (1.2) for $\alpha = 2$. But take into account that in the completely asymmetric stable case with $\alpha < 2$, the LIL does not hold for $|S_n|$ while LIL (1.2) holds. Nevertheless, the completely asymmetric stable laws are very similar to the normal law. They are only stable laws having one-sided exponential moments that yields this similarity. Hence, we concern with such stable laws in what follows.

Random variables with distributions from the domains of attractions of completely asymmetric stable laws appear in many theoretical and practical problems both. For example, X with the Pareto d.f. $F(x) = (-x)^{-\alpha}$, $x < -1$, belongs to the domain of attraction of the completely asymmetric stable law. The Pareto distribution is widely used in economics. It describes amounts of claims in some risk models of actuarial and financial mathematics, for example. In here, X is negative since every satisfied claim is a loss for an insurance company.

Further properties of trajectories of random walks are related with increments of sums S_n. Define functionals on these trajectories those we deal with.

Let $\{a_n\}$ be a non-decreasing sequence of natural numbers such that $1 \leq a_n \leq n$ for all n. Denote

$$U_n = \max_{0 \leq k \leq n-a_n} (S_{k+a_n} - S_k),$$
$$W_n = \max_{0 \leq k \leq n-a_n} \max_{1 \leq j \leq a_n} (S_{k+j} - S_k),$$
$$R_n = S_n - S_{n-a_n}, \quad T_n = S_{n+a_n} - S_n.$$

Maxima U_n and W_n have a clear sense in game settings when X_k is a gain of a player in k-th repetition of a game (negative gains are losses). Then U_n and W_n are the maximal gains of the player over successive series of repetitions of the game with length a_n and less than or equal to a_n, correspondingly. If n is regarded as time, then U_n and W_n are the maximal gains in time a_n and less than or equal to a_n. Increments R_n and T_n describe the evolution of a capital of the player as well. Thus, U_n, W_n, R_n and T_n are natural objects of the probability theory.

We see that

$$U_n = R_n = S_n \quad \text{and} \quad W_n = \max_{1 \leq k \leq n} S_k,$$

when $a_n = n$ for all n. If $a_n = 1$ for all n, then

$$U_n = W_n = \max_{1 \leq k \leq n} X_k.$$

Hence, the asymptotic behaviour of increments is close to that of sums S_n for large lengths a_n and that of maxima of i.i.d. random variables for small a_n. We built a universal theory (for all lengths a_n) which include results for sums, their increments and maxima of i.i.d. random variables.

Describe now the a.s. behaviour of increments of sums S_n.

First results of this type (the Erdős–Rényi and Shepp laws) have been proved in [Shepp (1964)] and [Erdős and Rényi (1970)]. They have established that for X with an exponential moment and $a_n = [c \log n]$, relation

$$\lim \frac{U_n}{a_n} = \limsup \frac{T_n}{a_n} = \gamma\left(\frac{1}{c}\right) \quad \text{a.s.} \tag{1.3}$$

holds. Here $\gamma(x)$ is the inverse function to the function of large deviations which we define in this section later. Now we postpone a discussion of properties of $\gamma(x)$ to Chapter 2. We only mention that $\gamma(x)$ depends on full distribution of X and determines this distribution sometimes. So, this result shows that trajectories of S_n can "remember" the distribution of X. This yields a variant of solution for the stochastic geyser problem: can Robinson Crusoe find a distribution of time between two successive eruptions of a geyser on his island? The answer is yes provided he knows the trajectory of S_n (see, [Csörgő and Révész (1981)] for details).

Note that [Csörgő (1979)] has proved a one-sided generalization of the Erdős–Rényi law. [Deheuvels and Devroye (1987)] have obtained (1.3) for $c \in (0, c_0)$ (see Chapter 2 for a definition of c_0). [Steinebach (1978)] and [Lynch (1983)] have proved that exponential moment conditions are necessary for (1.3).

[Mason (1989)] has obtained an extension of the Erdős–Rényi law for $a_n/\log n \to 0$. It turns out in this case that

$$\limsup \frac{U_n}{a_n \gamma(\log n/a_n)} = 1 \quad \text{a.s.} \tag{1.4}$$

and one can replace lim sup by lim for a class of distributions containing normal and exponential ones. One can also replace U_n by T_n in (1.4). This implies that if $a_n = m$ for all n and $P(X = -1) = P(X = 1) = 0.5$, then $\gamma(\infty) = 1$ and

$$\limsup \frac{X_{n+1} + \cdots + X_{n+m}}{m} = 1 \quad \text{a.s.}$$

So, the upper limit for m-period moving averages $(X_{n+1} + \cdots + X_{n+m})/m$ is not EX even for large, but fixed m. Note that moving averages are widely used in analysis of processes in economics on macro and micro levels both. In particular, trading terminals for Forex or stock markets allow to plot various moving averages over bars charts.

The case $a_n = 1$ for all n in (1.4) turns us to the asymptotic theory of extreme order statistics. Norming sequences in strong laws for maxima may depend on the distribution of X. If X has the standard normal distribution, then $\gamma(x) = \sqrt{2x}$ and

$$\lim \frac{1}{\sqrt{2 \log n}} \max_{1 \le k \le n} X_k = 1 \quad \text{a.s.}$$

For standard exponential X, the latter holds with $\log n$ instead of $\sqrt{2 \log n}$. From the other side, if $F(x) = 1 - 1/x$, $x \geq 1$, then for every increasing sequence $\{b_n\}$ of positive constants, one has

$$\limsup \frac{1}{b_n} \max_{1 \le k \le n} X_k = 0 \quad \text{or} \quad \infty \quad \text{a.s.}$$

and (1.4) fails. Further discussion on the asymptotic behaviour of extremes may be found in [Galambos (1978)], for example.

M.Csörgő and P.Révész have proved that if $EX = 0$, $EX^2 = 1$, X has the two-sided exponential moment and $a_n / \log n \to \infty$, then

$$\limsup \frac{U_n}{\sqrt{2a_n(\log(n/a_n) + \log\log n)}} = 1 \quad \text{a.s.} \tag{1.5}$$

and one can replace lim sup by lim in (1.5) provided $\log(n/a_n)/\log\log n \to \infty$ in addition (see [Csörgő and Révész (1981)]). For $a_n = n$, we have LIL (1.2). If $\log(n/a_n) \sim c \log\log n$, then we get the LIL with a certain constant instead of 1. [Csörgő and Révész (1981)] have also discussed weaker two-sided moment assumptions. Note that the moment assumption defines the minimal rate of the growth of a_n which relation (1.5) may hold for. One can find statistical applications in [Csörgő and Révész (1981)]. [Frolov (1990, 2000, 1998, 2002c, 2003b,d)] has obtained one-sided generalizations of the Csörgő–Révész laws.

Note that the behaviour of W_n is the same as that of U_n while for R_n and T_n, lim sup results hold only. Moreover, if increments of sums are replaced by moduli of increments in the definition of U_n, W_n, R_n and T_n, then the results will be the same.

The norming in (1.5) is the same for all distributions. Hence, large increments "forget" the distribution of X and "remember" its several moments only. This is a strong invariance while a strong non-invariance appears in (1.3).

One of the main goals of every science is a systematization. We see a series of results for sums S_n and their increments. We show that they may be written in one scheme and proved simultaneously.

More precisely, our main goal is to describe the a.s. asymptotic behaviour of U_n, W_n, R_n and T_n as $n \to \infty$. We will find sequences of positive numbers $\{b_n\}$ and sufficient and/or necessary conditions for either

$$\limsup \frac{U_n}{b_n} = 1 \quad \text{a.s.,} \tag{1.6}$$

or

$$\lim \frac{U_n}{b_n} = 1 \quad \text{a.s.} \tag{1.7}$$

Remember that all limits are taken as $n \to \infty$ if it is not pointed otherwise. Of course, we will consider analogues of (1.6) and (1.7) for W_n, R_n and T_n as well.

We see (1.6) and (1.7) in partial cases mentioned above. If $a_n = n$ and $b_n = nEX$ for all n with $EX > 0$, then relation (1.7) turns to SLLN (1.1). (Note that one can prove (1.1) for $EX > 0$ only. The general case easily follows by a centering of summands at $2EX$.) When $a_n = n$ and $b_n = \sqrt{2n \log\log n}$ for all n, relation (1.6) turns to LIL (1.2). For $a_n = [c \log n]$ and $b_n = a_n \gamma(1/c)$, relation (1.7) is (1.3). If $a_n = o(\log n)$ and $b_n = a_n \gamma(\log n / a_n)$, relation (1.6) coincides with (1.4). The case $a_n = 1$ for all n is in the results for extreme order statistics mentioned above.

One can prove strong laws by various methods.

M.Csörgő and P.Révész have used the strong approximation of sums by a Wiener process. This method is based on the following fact: on a certain probability space, one can define a sequence of sums $\{S_n\}$ of i.i.d. random variables and a Wiener process such that trajectories of sums and the process are closed enough. Then one can derive results for sums from results for the Wiener process.

If, for example, $EX = 0$, $EX^2 = 1$ and X has the two-sided exponential moment, then on a certain probability space, one can define $\{S_n\}$ and the standard Wiener process $W(t)$ such that

$$S_n - W(n) = O(\log n) \quad \text{a.s.} \tag{1.8}$$

This approximation is also referred in literature as Komlós–Major–Tushnády strong approximation. Note that the Erdős–Rényi law implies that O can not be replaced by o in the last relations. One can prove that if (1.8) holds with o instead of O, then X has the standard normal distribution.

Assume that a_T is a positive, non-decreasing function such that $a_T \leq T$ and T/a_T is non-decreasing. Suppose that w.p. 1, the process $W(t)$ has continuous trajectories. For increments of $W(t)$, M.Csörgő and P.Révész have proved that

$$\limsup_{T\to\infty} \frac{\sup_{0 \leq t \leq T - a_T} (W(t + a_T) - W(t))}{\sqrt{2a_T(\log(T/a_T) + \log\log T)}} = 1 \quad \text{a.s.} \tag{1.9}$$

One can replace lim sup by lim in (1.9) provided $\log(T/a_T)/\log\log T \to \infty$ as $T \to \infty$ (see [Csörgő and Révész (1981)]). Relation (1.9) holds for

$$\sup_{0 \leq t \leq T - a_T} \sup_{0 \leq s \leq a_T} (W(t+s) - W(t)),$$

$W(T) - W(T - a_T)$ and $W(T + a_T) - W(T)$ instead of the supremum in the left-hand side. For the last supremum, one can replace lim sup by lim in (1.9) when $\log(T/a_T)/\log\log T \to \infty$ as $T \to \infty$

Note that these results include the Erdős–Rényi law and the LIL for increments of the Wiener process. In particular, relation (1.9) for $a_T = T$ turns to the LIL

$$\limsup_{T\to\infty} \frac{W(T)}{\sqrt{2T\log\log T}} = 1 \quad \text{a.s.} \tag{1.10}$$

Moreover, for $a_T = 1$, we get a continuous analogue of the above result for maxima of i.i.d. standard normal random variables from the theory of extreme order statistics.

Since $-W(t)$ is the standard Wiener process as well, the above results hold for moduli of increments of the Wiener process.

Relations (1.9) and (1.8) imply (1.5) for $\{a_n\}$ such that $a_n/\log n \to \infty$. Further results for increments of sums follow in the same way.

For X without the exponential moment, the techniques is the same, but the accuracy in (1.8) decreases. This yields results for a more narrow range of $\{a_n\}$ as mentioned after relation (1.5).

The approximation like (1.8) does not work in general case. It can not be applied when $a_n = O(\log n)$ (the non-invariance) and $EX^2 = \infty$. For $EX^2 < \infty$, the accuracy of such approximation is not enough to provide optimal results under minimal assumptions on the left tail of the distribution of X. Hence, we use the method of an analysis of probabilities of large deviations or, briefly, the large deviations method. Note that the SLLN, the LIL, the Shepp law, the Erdős–Rényi law and its extension have been proved in this way.

To prove relations (1.6) and (1.7), the large deviations method will be applied as follows.

Since $R_n \le U_n \le W_n$, we derive the upper bound for W_n and the lower bounds for either R_n (in case of lim sup), or U_n (in case of lim).

For the upper bounds, we will estimate the probabilities $P_n = P(W_n \ge (1+\varepsilon)b_n)$ from above by probabilities of large deviations $P(S_{a_n} \ge (1+\varepsilon)b_n)$ with some coefficients. Appropriate bounds for the last probability imply that the series $\sum_k P_{n_k}$ converges for a special subsequence $\{n_k\}$ of natural numbers with $n_k \to \infty$ as $k \to \infty$. The Borel–Cantelli lemma yields

$$\limsup_{k\to\infty} \frac{W_{n_k}}{b_{n_k}} \le 1 + \varepsilon \quad \text{a.s.}$$

Monotonicity of W_n, regularity conditions on $\{b_n\}$ and limit passing as $\varepsilon \to 0$ imply the upper bound.

Turn to the lower bound in (1.6). If $a_n/n \to c < 1$, then we choose a sequence of natural numbers $\{n_k\}$ such that $n_k \to \infty$ as $k \to \infty$ and the events $\{R_{n_k} \geq (1-\varepsilon)b_{n_k}\}$ are independent. The divergence of the series from probabilities of these events is provided by appropriate asymptotics of probabilities of large deviations $P(S_{a_n} \geq (1-\varepsilon)b_n)$. Finally, the Borel–Cantelli lemma and limit passing as $\varepsilon \to 0$ yield the lower bound in (1.6). If $a_n/n \to 1$, then we put $n_k = c^k$ for all $k \in \mathbf{N}$ and prove that the series from probabilities of independent events $\{S_{n_k} - S_{n_{k+1}} \geq (1-\varepsilon)b_{n_k}\}$ converges. An application of a special analogue of the Borel–Cantelli lemma gives the result. Here, the asymptotics of large deviations of S_n is used as well.

Lower bound in (1.7) is derived in a similar way. Making use of asymptotics for probabilities of large deviations of sums S_n, we prove that for every $\varepsilon > 0$, the series $\sum_n P(U_n < (1-\varepsilon)b_n)$ converges. This and the Borel–Cantelli lemma imply the result.

Note that there are no specifications for $\{b_n\}$ in the above scheme. Hence, we first derive general results for arbitrary sequences $\{b_n\}$ and further find formulae for them from results on large deviations. Thus asymptotics of probabilities of large deviations of sums of i.i.d. random variables play a key role below.

We discuss the large deviations theory (LDT) in Chapter 2 that is of essential independent interest.

Probabilities of large deviations play an important role in statistics and various applications. Critical domains of statistical tests are often defined by large deviations of corresponding statistics. Many statistics coincide or may be approximated by sums of i.i.d. random variables. So, results for them easily follow.

We deal with the asymptotic behaviour of $\ln P(S_n \geq nx_n)$, where $\{x_n\}$ is a sequence of positive numbers such that $x_n = O(1)$.

In LDT, a significant place takes the large deviations function which we introduce as follows. Assume that X has the one-sided exponential moment. By the Tchebyshev inequality, we get

$$P(S_n \geq nx) \leq \mathrm{e}^{-n\zeta(x)},$$

where

$$\zeta(x) = \sup_{h \geq 0 \colon E\mathrm{e}^{hX} < \infty} \left\{hx - \log E\mathrm{e}^{hX}\right\}.$$

The function $\zeta(x)$ is called the large deviations function. We will show below that $\zeta(x)$ is well defined for $x \in [EX, A)$, where A will be specified later.

The function $\zeta(x)$ is the Legendre transform of logarithm of the moment generating function (m.g.f.) Ee^{hX}. Note that the Legendre transform is also used in mechanics and convex analysis, for example.

Probabilities of large deviations can be written in terms of conjugate distributions

$$\bar{F}(x) = \frac{1}{Ee^{hX}} \int_{-\infty}^{x} e^{hu} dP(X < u)$$

which have the two-sided exponential moments. We have

$$P(S_n \geq nx) = Ee^{hX} \int_{nx}^{\infty} e^{-hu} d\bar{F}(u).$$

Analysing the behaviour of the last integral, we arrive at results for large deviations.

Applying the method of conjugate distributions, we will prove that for X with the exponential moment, the relation

$$\ln P(S_n \geq nx) \sim -n\zeta(x)$$

holds for $X \in (EX, A)$. If we replace x by x_n in the last relations, the result may hold true provided $x_n \to 0$. Under additional assumptions on X and $\{x_n\}$, the asymptotics of $\zeta(x)$ at zero can be found. For example, if $EX = 0$, $EX^2 = 1$ and $x_n = o(\sqrt{n})$, then $\zeta(x_n) \sim x_n^2/2$. Note that $\zeta(x) = x^2/2$ for the standard normal law. Further we will show that similar results hold for X from domains of attraction of the normal law and the completely asymmetric stable laws with the exponent $\alpha \in (1, 2)$.

For X without the exponential moment, the method of truncations will be applied below. Truncated from above random variables have the exponential moment and, therefore, have an appropriate behaviour. They give the main part of the asymptotics of probabilities of large deviations while the asymptotics of rejected parts are negligible. The restrictions on $\{x_n\}$ will be stronger than those for X with the exponential moment.

From LDT, the large deviations function appears in further theory. Its inverse function $\gamma(x)$ or its variants determine the norming sequences in strong laws. Therefore, the SLLN, the LIL, the Erdős–Rényi law and its extension, the Shepp law and the Csörgő–Révész laws are partial cases of one universal law and we arrive at the universal theory of strong laws.

For X with the one-sided exponential moment, the norming is

$$b_n = a_n \gamma \left(\frac{\log(n/a_n) + \log \log n}{a_n} \right).$$

For $a_n = O(\log n)$, this is the norming from the Shepp law, the Erdős–Rényi law and its extension. Assume that $a_n/\log n \to 0$. For $EX > 0$, we get $b_n \sim a_n\gamma(0) = a_n EX$ that is the norming for SLLN. If $EX = 0$ and $EX^2 = 1$, then $\gamma(x) \sim \sqrt{2x}$ as $x \to 0$ and b_n is equivalent to the norming from the Csörgő–Révész laws. Similar results hold for domains of attractions of the normal law and the completely asymmetric stable laws with $\alpha \in (1,2)$.

For X without the exponential moment, $\gamma(x)$ in the above definition of b_n can be replaced by its variant for truncated from above random variables. Then we have the Csörgő–Révész laws. We also have their variants for the case of domains of attractions of the normal law and the completely asymmetric stable laws with $\alpha \in (1,2)$.

Note that that results will be derived under optimal moment conditions.

Besides sums of independent random variables, various classes of stochastic processes are of essential interest in probability, statistics and their applications. Below we deal with renewal processes and processes with independent increments.

For a sequence of positive i.i.d. random variables $\{Y_n\}$, the renewal process $N(t)$ is defined by

$$N(t) = \max\{n \geq 0 : R_n \leq t\},$$

where $R_0 = 0$, $R_n = Y_1 + Y_2 + \cdots + Y_n$. One of the most important case arises when Y_n has the exponential distribution and $N(t)$ is the Poisson process.

In particular, $N(t)$ describes the number of identical devices (electric bulbs, for instance) broken (and renewed) up to time t. Then $\{Y_n\}$ are random times of lives of the devices. In queuing theory, $N(t)$ may be the number of customers. In actuarial and financial mathematics, $N(t)$ may be the number of claims.

We choose renewal processes by two reasons. First, they are widely used in numerous applications. Second, a duality between R_n and $N(t)$ provides relatively simple passages from results for sums to results for renewal processes.

Indeed, we have

$$\{N(t) \geq n\} = \{R_n \leq t\} \tag{1.11}$$

for all $t > 0$ and natural n. This and the Kolmogorov SLLN for R_n yield the following SLLN for renewal processes:

$$\lim_{T\to\infty} \frac{N(T)}{T} = \frac{1}{a} \quad \text{a.s.}$$

provided $a = EY_1$ exists and it is positive. Note that $EN(1)$ is not equal to $1/a$ in the general case, but the equality holds for the Poisson process.

In the same way, duality (1.11) and the Hartman–Wintner LIL imply the next LIL for renewal processes:

$$\limsup_{T\to\infty} \frac{N(T) - (T/a)}{\sqrt{2T \log\log T}} = \frac{\sigma}{a^{3/2}} \quad \text{a.s.},$$

where $a = EY_1$ and $\sigma^2 = DY_1$. One can see that in the last relation, $N(t)$ is centered at $EN(T)$ for the Poisson process. The norming in this case is $\sqrt{DN(T)}$ provided we divide that relation by $\sigma/a^{3/2}$. This holds since the Poisson process is the homogeneous process with independent increments. Hence, the Kolmogorov SLLN and the Hartman–Wintner LIL hold for $\{N(n)\}$ which is the sequence of sums of i.i.d. random variables with the same distribution as $N(1)$.

Below, we use a more complicated techniques which gives results for increments of renewal processes. Besides an analogue of the supremum for the Wiener process from (1.9), we also investigate the following centered variants:

$$\sup_{0\le t\le T-a_T} \sup_{0\le s\le a_T} \left(N(t+s) - N(t) - \frac{s}{a}\right),$$

$$\sup_{0\le t\le T-a_T} (N(t+a_T) - N(t)) - \frac{a_T}{a}.$$

We construct a universal theory of strong laws for renewal processes. From the universal law, we derive the SLLN, the LIL, the Erdős–Rényi and the Csörgő–Révész laws for renewal processes. In there, the results depend on centering. We get the SLLN and the Erdős–Rényi law for non-centered processes and the LIL, the Erdős–Rényi law and the Csörgő–Révész laws for centered ones.

Homogeneous processes with independent increments are also of essential interest in probability, statistics and various applications.

The Wiener and Poisson processes are two first examples. The Wiener process or Brownian motion is a model for a motion of a particle randomly hitting by a large number of another ones. In probability, the Wiener process is often used as an approximation for random walks. Then results for functionals from random walks may be derived by applications of those for functionals from the Wiener process. We have above demonstrated this approach to sums of i.i.d. random variables. In statistics, this approach leads to results for empirical processes, for example.

Important processes with independent increments may be constructed from sums S_n and a Poisson process $N(t)$ (independent with sums) by

$$\xi(t) = S_{N(t)}.$$

This is the Compound Poisson process. Such processes are widely used in actuarial and financial mathematics as models for aggregate claim amounts of insurance companies for portfolios of insurance polices of the same type.

The behaviours of homogeneous processes with independent increments and sums of i.i.d. random variables are quite similar. For example, relations (1.5) and (1.9) turn each to other when we replace max by sup, sums S_n by the standard Wiener process $W(t)$, n by t and via versa. For these processes, some technical problems are related with continuous time, but increments are infinitely divisible. This and the importance of this class of processes are reasons for considerations below.

Note that we can easily define various suprema from the Poisson process and Compound Poisson process since their trajectories are step functions. The Wiener process has a modification which trajectories are continuous w.p. 1. For a general case, we consider stochastically continuous, homogeneous process $\xi(t)$ with independent increments such that $\xi(0) = 0$ a.s. and $E\xi(1) < \infty$. Then $\xi(t)$ has a modification which has trajectories from the space of cádlàg (right-continuous and having left limit at every point) functions on $[0, \infty)$ w.p. 1. We only concern with such modifications. Since $\xi(t)$ is right continuous and has left limits, we have no problems with definitions of considered functionals from trajectories of $\xi(t)$.

Stable processes are important examples of homogeneous processes with independent increments. We restrict our attention to such stable processes that $\xi(1)$ has a completely asymmetric stable distribution with the exponent $\alpha \in (1, 2)$. It turns out that these processes only have the one-sided exponential moment and finite mean. They are very similar to the Wiener process. Of course, they have modifications with trajectories from the space of cádlàg function (w.p. 1) while trajectories of the Wiener process are continuous w.p. 1. Nevertheless, analogues of a.s. results for the Wiener process hold for such stable processes as well. For example, if $E\xi(1) = 0$ and $\xi(1)$ has the m.g.f. $\mathrm{e}^{h^\alpha/\alpha}$ for $h > 0$, then the LIL is as follows:

$$\limsup_{T\to\infty} \frac{\xi(T)}{\lambda^{-\lambda} T^{1/\alpha} (\log\log T)^\lambda} = 1 \quad \text{a.s.},$$

where $\lambda = (\alpha - 1)/\alpha$. Putting formally $\alpha = 2$ in the last relation, we arrive at LIL (1.10) for the Wiener process. The nature of this relationship is clear because we can consider the standard normal distribution as "completely asymmetric" stable law with $\alpha = 2$. For $\alpha = 2$, the parameter of symmetry in the canonic representation of the stable c.f. is absent and we can choose it arbitrarily. Note that the LIL for $|\xi(t)|$ fails while the LIL for $|W(t)|$ holds.

We construct a universal theory of strong laws for homogeneous processes with independent increments. The Wiener process, the Compound Poisson process and the stable processes mentioned above are separately presented as important partial cases. From the universal strong laws, we derive the SLLN, the LIL, the Erdős–Rényi and the Csörgő–Révész laws for processes with independent increments.

Finally, we deal with the a.s. asymptotic behaviour of special functionals from trajectories of random walks.

We first state the problem in game settings. Suppose that the sequence of sums $\{S_n\}$ describes fluctuations of the capital of a player. We are interested in the behaviour of an aggregate amount of the gain of the player in series of games without losses. Formally, we arrive at the maximum

$$M_n' = \max_{0 \le k \le n-a_n} (S_{k+a_n} - S_k) I_{\{X_{k+1} \ge 0,\, X_{k+2} \ge 0,\, \dots,\, X_{k+a_n} \ge 0\}},$$

where I_B is the indicator of the event B. If the player pay a fee c for the participation in every repetition of the game, then we replace all zeros by c in the indicator of the event. Another interesting maximum is

$$M_n'' = \max_{0 \le k \le n-a_n} (S_{k+a_n} - S_k) I_{\{0 \le X_{k+1} \le X_{k+2} \le \dots \le X_{k+a_n}\}}.$$

It is the gain of a player in series with successively increasing gains. If there is a fee, then we replace 0 in the indicator by the fee c again.

Of course, the first question is as follows. What is the maximal length of the series in the indicators? If the sequence $\{a_n\}$ increases very fast, then it may happen that all indicators are zeros. Hence, the maxima will be zeros and the problem is trivial. It turns out that the maximal (random) length of considered series has the logarithmic order. Nevertheless, we will construct the corresponding universal theory and will show that all theorems mentioned above holds true for those maxima.

In the sequel, we deal with more general settings than before. We assume that we have an accompanying sequence of random variables which forms blocks in the indicators.

Let $(X, Y), (X_1, Y_1), (X_2, Y_2), \dots$ be a sequence of i.i.d. random vectors. Note that X and Y may be dependent and the case $X = Y$ is available as well. The sums of X's are denoted by S_n as before. Put

$$M_n = \max_{0 \le k \le n-j_n} (S_{k+[j_n]} - S_k) I_{\{u \le Y_{k+1} \le \dots \le Y_{k+[j_n]} \le v\}},$$

where u, v are fixed with $-\infty \leq u \leq v \leq +\infty$ and $\{j_n\}$ such that $1 \leq j_n \leq n$ for all n. Here $[x]$ is the integer part of the number x. Formally, j_n may also be a random variable, but we do not touch this case in there.

If $u = v = 1$ and $Y = I_{\{X \geq 0\}}$, then $M_n = M_n'$. If $u = 0$, $v = +\infty$ and $Y = X$, then $M_n = M_n''$.

When Y is a Bernoulli random variable with $P(Y = 1) = p = 1 - P(Y = 0)$, $p \in [0, 1]$, the series of Y's with $Y_{k+1} = Y_{k+2} \cdots = Y_{k+s} = 1$ is called a head run of length s. In this case, we mention M_n as the maximum of increments of sums over head runs. If Y is continuous, the series $Y_{k+1} \leq Y_{k+2} \leq \cdots \leq Y_{k+s}$ is called a monotone block (or an increasing run) of length s. In this case, M_n is the maximum of increments over monotone blocks. (For sake of brevity, we do not write further that sums are sums of i.i.d random variables and head runs and monotone blocks are in the accompanying sequence.) Note that there are investigations of another blocks of random variables in literature. Unfortunately, the relationship of large deviations of sums over blocks and sums of i.i.d. random variables is known for head runs and monotone blocks. So, there is a number of open questions.

It turns out that the length of the longest head run in $Y_1, Y_2, \ldots, Y_n$ is (asymptotically) $\log n / \log(1/p)$ a.s. while the length of the longest monotone block is $\log n / \log \log n$ a.s. Therefore, we are able to prove results for increments over head runs and monotone blocks simultaneously. At the same time, there is no sense to consider j_n greater than $\log n / \log(1/p)$ for head runs and $\log n / \log \log n$ for monotone blocks. Nevertheless, we will prove all the spectrum of the results as for sums of i.i.d. random variables. Below, we derive the SLLN, the LIL, the Erdős–Rényi and the Csörgő–Révész laws from the corresponding universal laws.

In the next section, we establish relationships between universal theorems and large deviations results for S_n those are basic for our theory below.

1.2 The Universal Strong Laws and the Large Deviations Method

We start this section with a natural choice for the length a_n of increments of sums S_n. Of course, we could consider any monotone sequences of natural numbers $\{a_n\}$, but this only yields more complicated statements of results. In what follows, a_n will be the integer part of the value of a continuous function at n.

Let $a(x)$ be a non-decreasing, continuous function such that $1 \le a(x) \le x$ for all $x \in \mathbf{R}$ and $x/a(x)$ is non-decreasing. For all $n \in \mathbf{N}$, put $a_n = [a(n)]$ and

$$\beta_n = \log \frac{n}{a_n} + \log\log(\max(n, 3)),$$

where $[\cdot]$ is the integer part of the number in brackets.

Our first result contains an a.s. upper bound for normalized maxima W_n.

Theorem 1.1. *Let $\{b_n\}$ be a sequence of positive constants. Assume that the following conditions hold:*

1) *the sequence $\{b_n\}$ is equivalent to some non-decreasing sequence and*

$$\lim_{\theta \searrow 1} \lim_{k \to \infty} \frac{b_{[\theta^{k+1}]}}{b_{[\theta^k]}} = 1. \tag{1.12}$$

2) $\sum_n P(X \ge cb_n) < \infty$ *for some* $c > 0$.

3) *For all small enough* $\varepsilon > 0$, *there exist* $\delta > 0$ *and* $H \ge 0$ *such that*

$$P\left(S_{[(1+\varepsilon)a_n]} \ge (1+\varepsilon)b_n\right) \le \mathrm{e}^{-(1+\delta)\beta_n} + Ha_nP(X \ge cb_n) \tag{1.13}$$

for all sufficiently large n.

4) *For all* $\varepsilon > 0$ *there exists* $q \in (0,1)$ *such that*

$$P\left(S_{[(1+\varepsilon)a_n]} \ge -\varepsilon b_n\right) \ge q \tag{1.14}$$

for all sufficiently large n.

Then

$$\limsup \frac{W_n}{b_n} \le 1 \qquad a.s. \tag{1.15}$$

One can replace W_n by T_n in the last relation.

If condition 3) holds with $H = 0$, then one can omit condition 2). If inequality (1.13) holds with $H = 0$ for all S_i, $1 \le i \le [(1+\varepsilon)a_n]$, and $\log a_n/\log n \to 0$, then one can omit conditions 2) and 4).

Proof. We need two lemmas. The first one is a variant of the Lévy inequality.

Lemma 1.1. *If $r, s \ge 0$, $q > 0$ and $P(S_i \ge -s) \ge q$ for all $i = 1, 2, \ldots, n$, then*

$$P\left(\max_{0 \le i < j \le n} (S_j - S_i) \ge r\right) \le q^{-2} P(S_n \ge r - 2s).$$

Proof. Put $M_i = \max_{i<j\le n} (S_j - S_i)$,

$$A_i = \{M_{n-1} < r, \ldots, M_{i+1} < r, M_i \ge r\} \quad \text{and} \quad D_i = \{S_i \ge -s\}$$

for $i = 0, 1, \ldots, n-1$. For all i, the independence of A_i and D_i yields that

$$qP(A_i) \le P(A_i)P(D_i) = P(A_iD_i).$$

Moreover,

$$A_iD_i \subset A = \left\{ \max_{1\le j\le n} S_j \ge r - s \right\}$$

for all i. Hence, we get

$$qP\left(\max_{0\le i<j\le n} (S_j - S_i) \ge r\right) = q\sum_{i=0}^{n-1} P(A_i) \le \sum_{i=0}^{n-1} P(A_iD_i)$$
$$= P\left(\bigcup_{i=0}^{n-1} A_iD_i\right) \le P(A).$$

Now put

$$A'_j = \{S_1 < r-s, \ldots, S_{j-1} < r-s, S_j \ge r-s\} \quad \text{and} \quad D'_j = \{S_n - S_j \ge -r\}$$

for $j = 1, 2, \ldots, n$. For all i, the independence of A'_i and D'_i implies that

$$qP(A'_i) \le P(A'_i)P(D'_i) = P(A'_iD'_i).$$

Moreover,

$$A'_iD'_i \subset \{S_n \ge r - 2s\}$$

for all i. Then we have

$$qP(A) = q\sum_{i=1}^{n} P(A'_i) \le \sum_{i=1}^{n} P(A'_iD'_i) = P\left(\bigcup_{i=1}^{n} A_iD_i\right) \le P(S_n \ge r - 2s).$$

The result follows. □

The next lemma implies that relation (1.14) holds for all S_i, $1 \le i \le [(1+\varepsilon)a_n]$, instead of $S_{[(1+\varepsilon)a_n]}$. We need the latter to apply Lemma 1.1.

Lemma 1.2. *Assume that $\{b_n\}$ is equivalent to some non-decreasing sequence. If condition 4) of Theorem 1.1 holds, then for every $\varepsilon > 0$ there exists $q > 0$ such that $P(S_i \ge -\varepsilon b_n) \ge q$ for all $i = 1, 2, \ldots, [(1+\varepsilon)a_n]$ and all sufficiently large n.*

Proof. Without loss of generality, we suppose that $\{b_n\}$ is non-decreasing itself and $a_n \to \infty$. Indeed, if $\{a_n\}$ is bounded then the conclusion of the lemma is obvious.

Put $I_n = [(1+\varepsilon)a_n]$. Since $x/a(x)$ is non-decreasing, we get $na(n-1) \geq (n-1)a(n)$. Hence, $0 \leq a(n) - a(n-1) \leq a(n)/n \leq 1$. It follows that $0 \leq I_n - I_{n-1} \leq (1+\varepsilon)(a_n - a_{n-1}) + 1 \leq (1+\varepsilon)(a(n) - a(n-1)) + 2 + \varepsilon \leq 3 + 2\varepsilon$. Then we have $0 \leq I_n - I_{n-1} \leq 3 + [2\varepsilon]$ for all n.

Take $\varepsilon > 0$. Then there exists $q > 0$ such that $P(S_{I_n} \geq -\varepsilon b_n) \geq q$ for all $n \geq N_0$.

Suppose that $n \geq N_0$ and $i \leq I_n$. If $i \geq I_{N_0}$, then there exists n_i such that $n_i \geq N_0$ and $I_{n_i} \leq i \leq I_{n_i} + m$, where $m = 2 + [2\varepsilon]$. Hence,

$$\begin{aligned}
P(S_i \geq -\varepsilon b_n) &= P(S_{I_{n_i}} + S_i - S_{I_{n_i}} \geq -\varepsilon b_n) \\
&\geq P(S_{I_{n_i}} \geq -\varepsilon b_n) P(S_i - S_{I_{n_i}} \geq 0) \\
&\geq P(S_{I_{n_i}} \geq -\varepsilon b_n)(P(X \geq 0))^m \\
&\geq P(S_{I_{n_i}} \geq -\varepsilon b_{n_i})(P(X \geq 0))^m \geq q(P(X \geq 0))^m.
\end{aligned}$$

If $i \leq I_{N_0}$, then $P(S_i \geq -\varepsilon b_n) \geq P(S_i \geq 0) \geq (P(X \geq 0))^{I_{N_0}} > 0$. Putting

$$q' = \min\{q(P(X \geq 0))^m, (P(X \geq 0))^{I_{N_0}}\},$$

we get the assertion of the lemma. □

Turn to the proof of Theorem 1.1. Without loss of generality, we suppose that $\{b_n\}$ is non-decreasing. We will check that (1.15) holds for

$$W_n^* = \max_{0 \leq k \leq n} \max_{1 \leq j \leq a_n} (S_{k+j} - S_k)$$

instead of W_n. It is clear that $W_n \leq W_n^*$ and $T_n \leq W_n^*$ for all n.

Assume first that conditions 1)–4) are satisfied and $H > 0$.

Take $\varepsilon > 0$ and denote

$$A_n(\varepsilon) = \{W_n^* \geq (1+\varepsilon)b_n\}.$$

By Lemmas 1.1 and 1.2 and relation (1.14), we have

$$\begin{aligned}
P_n &= P(A_n(3\varepsilon)) \\
&\leq P\left(\bigcup_{j=1}^{[n/(\varepsilon a_n)]+1} \left\{\max_{(j-1)\varepsilon a_n \leq m \leq j\varepsilon a_n} \max_{1 \leq k \leq a_n} (S_{m+k} - S_m) \geq (1+3\varepsilon)b_n\right\}\right) \\
&\leq \sum_{j=1}^{[n/(\varepsilon a_n)]+1} P\left(\max_{(j-1)\varepsilon a_n \leq m \leq j\varepsilon a_n} \max_{1 \leq k \leq a_n} (S_{m+k} - S_m) \geq (1+3\varepsilon)b_n\right) \\
&\leq \frac{(1+\varepsilon)n}{\varepsilon a_n} P\left(\max_{0 \leq m \leq \varepsilon a_n} \max_{1 \leq k \leq a_n} (S_{m+k} - S_m) \geq (1+3\varepsilon)b_n\right) \\
&\leq \frac{(1+\varepsilon)n}{\varepsilon a_n q^2} P\left(S_{[(1+\varepsilon)a_n]} \geq (1+\varepsilon)b_n\right) \qquad (1.16)
\end{aligned}$$

for all sufficiently large n.

Take $\theta > 1$. For $k \in \mathbf{N}$, put

$$n_k = \min \Big\{ n : \theta^{k-1} < n \le \theta^k,$$

$$nP(X \ge cb_n) = \min_{\theta^{k-1} < m \le \theta^k} mP(X \ge cb_m) \Big\}. \qquad (1.17)$$

Then

$$\sum_{\theta^{k-1} < n \le \theta^k} P(X \ge cb_n)$$

$$\ge n_k P(X \ge cb_{n_k}) \sum_{\theta^{k-1} < n \le \theta^k} \frac{1}{n}$$

$$\ge n_k P(X \ge cb_{n_k}) \frac{\theta - 1}{\theta}.$$

It follows that

$$\frac{\theta - 1}{\theta} \sum_{k=1}^{\infty} n_k P(X \ge cb_{n_k}) \le \sum_{k=1}^{\infty} \sum_{\theta^{k-1} < n \le \theta^k} P(X \ge cb_n) = \sum_{n=1}^{\infty} P(X \ge cb_n).$$

Condition 2) implies that the series $\sum_{k=1}^{\infty} n_k P(X \ge cb_{n_k})$ converges.

By (1.13) and (1.16), we have

$$P_{n_k} \le C_1 \frac{n_k}{a_{n_k}} \exp\{-(1+\delta)\beta_{n_k}\} + C_2 n_k P(X \ge cb_{n_k})$$

for all sufficiently large k. Making use of the definition of β_n, we get

$$P_{n_k} \le C_1 (\log n_k)^{-(1+\delta)} + C_2 n_k P(X \ge cb_{n_k}) \qquad (1.18)$$

for all sufficiently large k. It yields that the series $\sum_{k=1}^{\infty} P_{n_k}$ converges. By the Borel–Cantelli lemma, we obtain

$$\limsup_{k \to \infty} \frac{W_{n_k}^*}{b_{n_k}} \le 1 + 3\varepsilon \quad \text{a.s.} \qquad (1.19)$$

For all n with $n_{k-1} < n \le n_k$, the following inequalities hold:

$$\frac{W_n^*}{b_n} \le \frac{W_{n_k}^*}{b_{n_{k-1}}} = \frac{W_{n_k}^*}{b_{n_k}} \frac{b_{n_k}}{b_{n_{k-1}}} \le \frac{W_{n_k}^*}{b_{n_k}} \frac{b_{[\theta^k]}}{b_{[\theta^{k-2}]}}. \qquad (1.20)$$

Condition (1.12) implies that for every $\tau > 0$, the inequality $b_{[\theta^k]}/b_{[\theta^{k-2}]} < (1+\tau)^2$ holds for all θ close to 1 and all sufficiently large k. Therefore,

$$\limsup \frac{W_n^*}{b_n} \le (1 + 3\varepsilon)(1+\tau)^2 \quad \text{a.s.}$$

Passing to the limit as $\varepsilon \downarrow 0$ and $\tau \downarrow 0$ in the last inequality, we obtain (1.15) for W_n^*.

Assume now that $H = 0$. Then $C_2 = 0$ in (1.18). Condition 2) has been only used to prove the convergence of $\sum_k n_k P(X \geq b_{n_k})$ that is not needed for $C_2 = 0$. Hence, condition 2) is not supposed any longer. The above proof may be repeated with $n_k = [\theta^k]$ instead of n_k from (1.17). Details are omitted.

Let us turn to the last case. Assume that inequality (1.13) holds with $H = 0$ for all S_i, $i \leq [(1+\varepsilon)a_n]$, and $\log a_n / \log n \to 0$. Then

$$P(A_n(\varepsilon)) \leq n \sum_{i=1}^{a_n} P(S_i \geq (1+\varepsilon)b_n) \leq n \sum_{i=1}^{a_n} \mathrm{e}^{-(1+\delta)\beta_n}$$
$$= n^{-\delta} a_n^{2+\delta} (\log n)^{-(1+\delta)}.$$

This yields that the inequality

$$P(A_n(\varepsilon)) \leq (\log n)^{-(1+\delta)}$$

holds for all sufficiently large n. Put $n_k = [\theta^k]$. It is clear that the series $\sum_{k=1}^{\infty} P(A_{n_k}(\varepsilon))$ converges. The rest of the proof coincides with that of the above proof. We omit details.

Note that the case $a_n = const$ for all n is included here as well. □

Remark 1.1. If relation (1.14) holds with S_i, $1 \leq i \leq [(1+\varepsilon)a_n]$, instead of $S_{[(1+\varepsilon)a_n]}$ for all sufficiently large n, then one can replace condition 1) of Theorem 1.1 by

$$\lim_{\theta \searrow 1} \limsup_{k\to\infty} \frac{b_{n_k}}{m_k} = 1, \tag{1.21}$$

where $m_k = \min_{n_{k-1} \leq n \leq n_k} b_n$ and n_k is from (1.17). Condition 1) of Theorem 1.1 is sufficient for (1.21).

Proof. Besides Lemma 1.2, condition 1) has been only used in (1.20) and the bound for $b_{[\theta^k]}/b_{[\theta^{k-2}]}$ later. If condition (1.21) is fulfilled, then for every $\tau > 0$, the inequality $b_{n_k}/m_k < 1 + \tau$ holds for all θ close enough to 1 and all sufficiently large k. Therefore, for all n with $n_{k-1} < n \leq n_k$, the inequalities

$$\frac{W_n^*}{b_n} \leq \frac{W_n^*}{b_{n_k}} \frac{b_{n_k}}{m_k} \leq \frac{W_n^*}{b_{n_k}}(1+\tau)$$

hold for all sufficiently large k. It follows that

$$\limsup \frac{W_n^*}{b_n} \leq (1+2\varepsilon)(1+\tau) \quad \text{a.s.},$$

from which we get (1.15) for W_n^* by limit passings as $\varepsilon \downarrow 0$ and $\tau \downarrow 0$.

One can easily check that 1) is sufficient for (1.21). We omit details. □

We have used inequality (1.13) for the subsequence $\{n_k\}$ in the proof of Theorem 1.1. It follows the next result.

Remark 1.2. On can replace inequality (1.13) in Theorem 1.1 by

$$P\left(S_{[(1+\varepsilon)a_{n_k}]} \geq (1+\varepsilon)b_{n_k}\right) \leq \mathrm{e}^{-(1+\delta)\beta_{n_k}} + Ha_{n_k}P(X \geq cb_{n_k}), \quad (1.22)$$

where n_k is from formula (1.17).

Our next result contains a.s. lower bounds for upper limits of normalized increments. This result and Theorem 1.1 allow to find the maximal rate of the growth of increments of sums of i.i.d. random variables.

Theorem 1.2. *Assume that for every $\varepsilon > 0$ there exists $\tau > 0$ such that*

$$P(S_{a_n} \geq (1-\varepsilon)b_n) \geq \mathrm{e}^{-(1-\tau)\beta_n} \quad (1.23)$$

for all sufficiently large n. If $a_n/n \to 1$, then assume in addition that conditions 1) and 4) of Theorem 1.1 are satisfied.

Then

$$\limsup \frac{R_n}{b_n} \geq 1 \quad a.s. \quad (1.24)$$

One can replace R_n by T_n in the last relation.

Proof. By (1.23), we have

$$P_n = P(R_n \geq (1-\varepsilon)b_n) \geq \mathrm{e}^{-(1-\tau)\beta_n} \geq \frac{a_n}{n}(\log n)^{-(1-\tau)} \quad (1.25)$$

for all sufficiently large n.

Assume first that $a_n/n \to \varrho \in [0,1)$. Choose $\delta \in (\varrho, 1)$ and a natural number N such that inequalities $(a_n+1)/n < \delta$ and (1.25) hold for all $n > N$. Put $n_1 = N$ and

$$n_{k+1} = \min\{n : n > n_k,\, n - a_n \geq n_k\}$$

for $k \in \mathbf{N}$. Making use of the inequality $a_n \geq (a_n+1)/2$, we get

$$\sum_{k=2}^{m} P_{n_k} \geq \frac{1}{2}(\log n_m)^{-1+\tau} \sum_{k=2}^{m} \frac{a_{n_k}+1}{n_k}.$$

It yields from the inequalities $n_k - a_{n_k} - 1 \le n_k - a_{n_k-1} - 1 < n_{k-1}$ and $-\log(1-x) \le C_\delta x$ for $x \in (0,\delta)$ that

$$C_\delta \sum_{k=2}^{m} \frac{a_{n_k}+1}{n_k} \ge \sum_{k=2}^{m} \log \frac{n_k}{n_k - a_{n_k} - 1} \ge \sum_{k=2}^{m} \log \frac{n_k}{n_{k-1}} = \log \frac{n_m}{n_1}.$$

It follows that

$$\sum_{k=2}^{m} P_{n_k} \ge C(\log n_m)^\tau,$$

and the series $\sum_k P_{n_k}$ diverges. Using the independence of the events $\{R_{n_k} \ge (1-\varepsilon)b_{n_k}\}$, we arrive at the conclusion of the theorem from the Borel–Cantelli lemma.

Suppose now that $a_n/n \to 1$. Put $n_k = [\theta^k]$ for $k \in \mathbf{N}$, where $\theta > 1$. We will check that relation

$$\limsup_{k\to\infty} \frac{R_{n_k}}{b_{n_k}} \le a < 1 \quad \text{a.s.} \tag{1.26}$$

does not hold. To this end we will use the next result.

Lemma 1.3. *Let $\{A_n\}$ and $\{D_n\}$ be two sequences of events such that for every n the following pairs of events are independent: A_n and D_n, A_n and $D_n\overline{A_{n-1}D_{n-1}}$, A_n and $D_n\overline{A_{n-1}D_{n-1}}\ \overline{A_{n-2}D_{n-2}}$, ...*

If the series $\sum_n P(A_n)$ diverges, then

$$P(A_nD_n\ i.o.) \ge \liminf P(D_n).$$

Proof. Put $p = \liminf P(D_n)$ and suppose that $p > 0$. For all $n \ge 1$ and $j \ge 0$, put

$$U_n^j = \bigcup_{k=n}^{n+j} A_kD_k, \quad U_n^\infty = \bigcup_{k=n}^{\infty} A_kD_k, \quad I_n^j = \overline{U_n^j} = \bigcap_{k=n}^{n+j} \overline{A_kD_k}.$$

Assume that the conclusion of the lemma does not hold. Then

$$q = P(A_nD_n\,\text{i.o.}) = \lim_{n\to\infty} P(U_n^\infty) < p.$$

Take $\delta = (p-q)/3$. Then there exists N such that for every $n \ge N$, the inequalities $P(D_n) \ge p - \delta$ and $P(U_n^j) \le q + \delta$ hold for all j .

Suppose that $n \ge N$. Then

$$P(D_{n+j}I_n^{j-1}) = P(D_{n+j}) - P(D_{n+j}U_n^{j-1}) \ge \delta$$

and

$$P(A_{n+j}D_{n+j}I_n^{j-1}) = P(A_{n+j})P(D_{n+j}I_n^{j-1}) \ge \delta P(A_{n+j})$$

for all $j \geq 1$. It follows that

$$P(U_n^\infty) = P(A_n D_n) + \sum_{j=1}^{\infty} P(A_{n+j} D_{n+j} I_n^{j-1}) \geq \delta \sum_{k=n+1}^{\infty} P(A_k) = \infty.$$

This contradiction shows that the above assumption is not true. □

Let us consider the events

$$A_k = \{S_{n_k} - S_{n_{k-1}} > (1+\varepsilon)ab_{n_k}\}$$

and

$$D_k = \left\{S_{n_{k-1}} - S_{n_k - a_{n_k}} > -\varepsilon ab_{n_k}\right\}$$

for $k = 2, 3, \ldots$.

The relation $n_{k-1} - n_k + a_{n_k} \sim n_{k-1}$ as $k \to \infty$ implies that $n_{k-1} - n_k + a_{n_k} \leq n_k$ for all sufficiently large k. Condition 4) and Lemma 1.2 yield that $P(D_k) \geq q$ for all sufficiently large k.

Choose m_k such that $a_{m_k} = n_k - n_{k-1}$. This is possible because the set of values of a_n (as a function of n) coincides with $\mathbf{N}$. The relation $a_{n_k} \sim n_k \sim (n_k - n_{k-1})\theta/(\theta - 1)$ as $k \to \infty$ yields that $n_k \sim m_k\theta/(\theta-1)$ as $k \to \infty$.

Condition 1) implies that for every $\delta > 0$, there exists $\varrho_0 > 1$ such that $b_{[\varrho t]} \leq (1+\delta) b_{[t]}$ for all $1 < \varrho < \varrho_0$ and all sufficiently large t. Choosing large θ with $\varrho = \theta/(\theta-1) < \varrho_0$, we conclude that $b_{n_k} \leq b_{[\varrho^2 m_k]} \leq (1+\varepsilon)^2 b_{m_k}$ for all sufficiently large k.

It follows that

$$\begin{aligned} P(A_k) &= P(S_{n_k - n_{k-1}} > (1+\varepsilon)ab_{n_k}) = P(S_{a_{m_k}} > (1+\varepsilon)ab_{n_k}) \\ &\geq P(S_{a_{m_k}} > (1+\varepsilon)^3 ab_{m_k}) \end{aligned}$$

for all sufficiently large k. Choose small ε such that $(1+\varepsilon)^3 a < 1$. By (1.23), we get

$$P(A_k) \geq \exp\{-(1-\tau)\beta_{n_k}\} \geq k^{-(1-\tau)}$$

for all sufficiently large k. Then the series $\sum_k P(A_k)$ diverges.

By Lemma 1.3, we have

$$P(A_k D_k \text{ i.o.}) \geq q > 0.$$

From the other hand, relation (1.26) implies that

$$P(A_k D_k \text{ i.o.}) = 0.$$

This contradiction shows that (1.26) does not hold and the theorem is proved for R_n.

The proof for T_n follows the same way. By (1.23), we have

$$Q_n = P(T_n \geq (1-\varepsilon)b_n) \geq \mathrm{e}^{-(1-\tau)\beta_n} \tag{1.27}$$

for all sufficiently large n.

Assume first that $a_n/n \to \varrho \in [0,1)$. Choose $\delta \in (\varrho, 1)$ and a natural N such that inequalities $(a_n+1)/n < \delta$ and (1.27) hold for all $n > N$. Put $n_1 = N$ and

$$n_{k+1} = \min\{n : n > n_k, n \geq n_k + a_{n_k}\}$$

for $k \in \mathbf{N}$. In the same way as before, we obtain

$$\sum_{k=2}^{m} Q_{n_k} \geq C(\log n_m)^{\tau}.$$

Then the series $\sum_k Q_{n_k}$ diverges and the events $\{T_{n_k} \geq (1-\varepsilon)b_{n_k}\}$ are independent. The desired assertion follows from the Borel–Cantelli lemma.

Suppose now that $a_n/n \to 1$. Denote $n_k = [\theta^k]$ for $k \in \mathbf{N}$, where $\theta > 1$. We will prove that relation

$$\limsup_{k\to\infty} \frac{T_{n_k}}{b_{n_k}} \leq a < 1 \quad \text{a.s.} \tag{1.28}$$

does not hold.

For $k = 2, 3, \ldots$, consider the events

$$A'_k = \left\{S_{n_k+a_{n_k}} - S_{n_{k-1}+a_{n_k}} > (1+\varepsilon)ab_{n_k}\right\}$$

and

$$D'_k = \left\{S_{n_{k-1}+a_{n_k}} - S_{n_k} > -\varepsilon ab_{n_k}\right\}.$$

Condition 4) and Lemma 1.2 imply that $P(D'_k) \geq q$ for all sufficiently large k.

In the same way as before, it follows from relation (1.23) that the series $\sum_k P(A'_k)$ diverges.

Applying Lemma 1.3, we conclude again that relation (1.28) does not hold. The theorem is now proved for T_n as well. □

If the sequence $\{a_n\}$ increases slow enough, then the behaviour of U_n and W_n becomes more stable that follows from the next result.

Theorem 1.3. *If the conditions of Theorem 1.2 hold and* $\log\log n = o(\log(n/a_n))$, *then*

$$\liminf \frac{U_n}{b_n} \geq 1 \quad a.s. \tag{1.29}$$

Proof. Take $\varepsilon > 0$ and put

$$A_n = \{U_n \le (1-\varepsilon)b_n\} .$$

Making use of (1.23), $n/a_n \to \infty$ and the inequality $1 - x \le e^{-x}$, $x > 0$, we get

$$P(A_n) \le P\left(\bigcap_{m=0}^{[n/a_n]-1} \{S_{(m+1)a_n} - S_{ma_n} \le (1-\varepsilon)b_n\}\right)$$

$$= (P(S_{a_n} \le (1-\varepsilon)b_n))^{[n/a_n]-1} \le \left(1 - \mathrm{e}^{-(1-\tau)\beta_n}\right)^{[n/a_n]-1}$$

$$= \left(1 - \left(\frac{a_n}{n\log n}\right)^{1-\tau}\right)^{[n/a_n]-1} \le \left(1 - \left(\frac{a_n}{n\log n}\right)^{1-\tau}\right)^{n/(2a_n)}$$

$$\le \exp\left\{-\frac{1}{2}\left(\frac{n}{a_n}\right)^{\tau} (\log n)^{-1+\tau}\right\}$$

for all sufficiently large n.

It follows from $\log\log n = o(\log(n/a_n))$ that the inequality $n/a_n \ge (\log n)^{2/\tau}$ holds for all sufficiently large n. Then

$$P(A_n) \le \exp\left\{-\frac{1}{2}(\log n)^{1+\tau}\right\} \le n^{-2}$$

for all sufficiently large n. This implies that the series $\sum_n P(A_n)$ converges. The Borel–Cantelli lemma yields (1.29). □

Theorems 1.1–1.3 and the inequalities $R_n \le U_n \le W_n$ imply the following result.

Theorem 1.4. *If the conditions of Theorems 1.1 and 1.2 hold, then*

$$\limsup \frac{W_n}{b_n} = \limsup \frac{U_n}{b_n} = \limsup \frac{R_n}{b_n} = \limsup \frac{T_n}{b_n} = 1 \quad a.s.$$

If, in addition, $\log\log n = o(\log(n/a_n))$, *then*

$$\lim \frac{W_n}{b_n} = \lim \frac{U_n}{b_n} = 1 \quad a.s.$$

Theorems 1.1–1.4 work for various sequences $\{a_n\}$. Therefore, they are the universal strong laws. Note that other authors may use this term in another senses. For example, results under minimal moment assumptions on X may be called universal as well.

The assumptions of Theorems 1.1–1.4 are very general. They illustrate tools that we need to prove strong limit theorems for increments. Relations

(1.13) and (1.23) are bounds for probabilities of large deviations for sums S_n. Inequality (1.14) is a kind of the Lévy inequality. Condition (1.12) is a regularity condition for norming sequence $\{b_n\}$ that will be specified later.

In the sequel, we will first study the asymptotic behaviour of probabilities of large deviations for S_n. This will allow us to find a formula for b_n and to obtain further universal results under various one-sided moment assumptions. Then we derive corollaries of the universal strong laws that include the results mentioned above. We will also prove that the moment assumptions are optimal.

Chapter 2

Large Deviations for Sums of Independent Random Variables

Abstract. We discuss the method of conjugate distributions, properties of related functions (the moment generating function and its logarithmic derivatives, the large deviations function and its inverse) and a classification of probability distributions with exponential moments. We derive asymptotics of these functions at zero for random variables from domains of attractions of the normal law and completely asymmetric stable laws with index from $(1, 2)$. Then we find logarithmic asymptotics of large deviations for such random variables. For infinite exponential moments, we use truncations. Finally, we obtain large deviations results associated with the classification of distributions.

2.1 Probabilities of Large Deviations

Let $\{X_n\}$ be a sequence of i.i.d. random variables and $\{x_n\}$ be a sequence of positive real numbers such that $x_n \to \infty$ as $n \to \infty$. For all $n \in \mathbf{N}$, put

$$S_n = X_1 + X_2 + \cdots + X_n.$$

Probabilities $P(S_n \geq x_n)$ are called probabilities of large deviations. We will assume that the sequences $\{x_n\}$ increase fast enough. For example, if $EX_1 = 0$ and $EX_1^2 = 1$, then by the central limit theorem, asymptotic behaviours of $P(S_n \geq x_n)$ and $1 - \Phi(x_n/\sqrt{n})$ coincide for $x_n = O(\sqrt{n})$, where $\Phi(x)$ is the standard normal d.f. Hence, in this case, we will only deal with the sequences $\{x_n\}$ such that $x_n/\sqrt{n} \to \infty$.

Investigations of the asymptotic behaviour of probabilities of large deviations are of essential interest in probability theory and its applications. We mentioned above that the method of analysis of large deviations probabilities allows to prove various strong limit theorems under optimal one-sided moment assumptions. We will use this method to derive the universal laws for sums of independent random variables. To this end, we only need asymptotics of functions of the large deviations theory (LDT) that we will introduce below. At the same time, these asymptotics may be used to derive logarithmic asymptotics of probabilities of large deviations which is of independent interest. We will start this chapter with a method of investigations of the large deviations. Further, we will describe the behaviour of the functions of the LDT and their relationships with the distributions of summands. Moreover, we will find conditions under which logarithms of probabilities of large deviations have the same asymptotics as logarithms of tails of the standard normal distribution or completely asymmetric stable laws with index $\alpha \in (1, 2)$.

2.2 The Method of Conjugate Distributions

In this section, we derive a method of investigations of asymptotic behaviour for large deviation probabilities of sums of independent random variables.

Start with a single random variable X.

Let X be a non-degenerate random variable such that $EX \geq 0$ and

$$h_0 = \sup\{h : Ee^{hX} < \infty\} > 0. \tag{2.1}$$

The function $\varphi(h) = Ee^{hX}$, $0 \leq h < h_0$, is called the moment generating function (m.g.f.). Condition (2.1) provides the existence of the function $\varphi(h)$ in some interval with left point at 0. It is called the one-sided Cramér condition. Note that it imposes no assumptions on the left tail of the distribution of X and, consequently, one may consider the random variable X without the second moment as well.

First and second logarithmic derivatives of $\varphi(h)$ play an important role in what follows. For $0 \leq h < h_0$, put

$$m(h) = (\log \varphi(h))' = \frac{\varphi'(h)}{\varphi(h)},$$

$$\sigma^2(h) = (\log \varphi(h))'' = \frac{\varphi''(h)}{\varphi(h)} - \left(\frac{\varphi'(h)}{\varphi(h)}\right)^2.$$

The method of conjugate distributions is based on a passing from the distribution of X to a distribution having all moments.

Put $F(x) = P(X < x)$ for $x \in \mathbf{R}$. The distribution, corresponding to the d.f.

$$\bar{F}(x) = \frac{1}{\varphi(h)} \int_{-\infty}^{x} e^{hu} dF(u),$$

is called a conjugate distribution. Here $h \in (0, h_0)$ is a parameter which we can vary appropriately.

Let $\bar{X}$ be a random variable with the d.f. $\bar{F}(x)$. We have

$$m(h) = \frac{1}{\varphi(h)} \int_{-\infty}^{\infty} u e^{hu} dF(u) = \int_{-\infty}^{\infty} u d\bar{F}(u) = E\bar{X},$$

$$\sigma^2(h) = \frac{1}{\varphi(h)} \int_{-\infty}^{\infty} u^2 e^{hu} dF(u) - m^2(h) = E\bar{X}^2 - (E\bar{X})^2 = D\bar{X},$$

for $0 \le h < h_0$. One can easily find all moments of the conjugate random variable $\bar{X}$.

This yields the following properties of the functions $\varphi(h)$, $m(h)$ and $\sigma^2(h)$:

1) $m(h)$ is a continuous, strictly increasing function with $m(0) = EX$;

2) $\sigma^2(h)$ is continuous, $\sigma^2(0) = DX$ (if the variation is finite) and $\sigma^2(h) > 0$;

3) $\varphi(0) = 1$, $\varphi(h)$ is a convex function and $\log \varphi(h)$ has derivatives of all orders in the interval $0 < h < h_0$.

Further, by the definition of the d.f. $\bar{F}(x)$, we have

$$P(X \ge x) = \varphi(h) \int_{x}^{\infty} e^{-hu} d\bar{F}(u).$$

This relation allows us to investigate the behaviour of the tail of the distribution of X and, therefore, that of large deviation probabilities. The corresponding method is called the method of conjugate distributions.

Lemma 2.1. *Let X be a non-degenerate random variable such that $EX \ge 0$ and $h_0 > 0$. Then for all $h \in (0, h_0)$, the following inequalities hold*

$$P(X \ge m(h)) \le e^{-f(h)},$$
$$P(X \ge m(h) - 2\sigma(h)) \ge \frac{3}{4} e^{-f(h) - 2h\sigma(h)},$$

where $f(h) = hm(h) - \log \varphi(h)$.

Proof. The first inequality follows from the Tchebyshev inequality. Prove the second one. Put $Z = (\bar{X} - m(h))/\sigma(h)$, $V(x) = P(Z < x)$. Then

$$P(X \geq m(h) - 2\sigma(h)) = \varphi(h) \int\limits_{m(h)-2\sigma(h)}^{\infty} e^{-hu} d\bar{F}(u)$$

$$= \varphi(h) \int\limits_{-2}^{\infty} e^{-h(m(h)+y\sigma(h))} dV(y) = e^{-f(h)} \int\limits_{-2}^{\infty} e^{-hy\sigma(h)} dV(y)$$

$$\geq e^{-f(h)} \int\limits_{-2}^{2} e^{-hy\sigma(h)} dV(y) \geq e^{-f(h)-2h\sigma(h)} P(|Z| \leq 2)$$

$$\geq e^{-f(h)-2h\sigma(h)} \left(1 - \frac{EZ^2}{4}\right) = \frac{3}{4} e^{-f(h)-2h\sigma(h)}.$$

In the last inequality, we have used the Tchebyshev inequality again. □

It follows from the properties of the functions $\varphi(h)$, $m(h)$ and $\sigma^2(h)$ mentioned above that the function $f(h)$ is continuous, strictly increasing and $f(0) = 0$. The second property follows from the relation $f'(h) = h\sigma^2(h)$.

To apply Lemma 2.1 for $X = S_n$, where S_n is a sum of i.i.d. random variables, we only need to find the m.g.f. and its logarithmic derivatives for S_n. It is clear that $Ee^{hS_n} = (\varphi(h))^n$. It follows that $m(h)$, $f(h)$ and $\sigma^2(h)$ have to be replaced in Lemma 2.1 by $nm(h)$, $nf(h)$ and $n\sigma^2(h)$ correspondingly. Then we get the following result.

Lemma 2.2. *Let $X, X_1, X_2, \ldots, X_n$ be i.i.d. non-degenerate random variables such that $EX \geq 0$ and $h_0 > 0$. Put $S_n = X_1 + X_2 + \cdots + X_n$. Then the inequalities*

$$P(S_n \geq nm(h)) \leq e^{-nf(h)},$$
$$P(S_n \geq nm(h) - 2\sqrt{n}\sigma(h)) \geq \frac{3}{4} e^{-nf(h)-2h\sqrt{n}\sigma(h)}$$

hold for all $h \in (0, h_0)$.

We will see below that logarithmic asymptotics of left-hand and right-hand sides of the inequalities in Lemma 2.2 coincide under certain conditions. This will allow us to find the logarithmic asymptotics of large deviation probabilities for sums of i.i.d. random variables.

2.3 Completely Asymmetric Stable Laws with Exponent $\alpha > 1$

We consider completely asymmetric stable laws F_α with the characteristic function (c.f.)

$$\psi(t) = \exp\left\{-c|t|^\alpha \left(1 + i\frac{t}{|t|}\tan\frac{\pi\alpha}{2}\right)\right\}, \tag{2.2}$$

where $c = -\cos(\pi\alpha/2)/\alpha$ and $\alpha \in (1, 2]$. For $\alpha = 2$, it is the standard normal c.f.

Such stable laws have densities, zero means and finite exponential moments. Indeed, for all real t, we have

$$\psi(t) = \exp\left\{\frac{|t|^\alpha}{\alpha}\left(\cos\frac{\pi\alpha}{2} + i\frac{t}{|t|}\sin\frac{\pi\alpha}{2}\right)\right\}$$
$$= \exp\left\{\frac{|t|^\alpha}{\alpha} e^{i\frac{t}{|t|}\frac{\pi\alpha}{2}}\right\} = \exp\left\{\frac{(it)^\alpha}{\alpha}\right\}.$$

An analytic extension of $\psi(t)$ in the complex half-plane $\{z : \mathrm{Re} z \geq 0\}$ is the function $\psi(z) = \exp\{z^\alpha/\alpha\}$. Hence, the distribution with the c.f. (2.2) has the m.g.f.

$$\varphi(h) = \exp\left\{\frac{h^\alpha}{\alpha}\right\} \quad \text{for} \quad h > 0.$$

It follows that

$$m(h) = h^{\alpha-1}, \quad \sigma^2(h) = (\alpha - 1)h^{\alpha-2}, \quad f(h) = \frac{\alpha - 1}{\alpha} h^\alpha$$

for all $h > 0$. Here $\alpha = 2$ corresponds to the standard normal law when one can check by direct calculations that $\varphi(h) = h^2/2$ for all real h. Of course, the standard normal law is the only completely asymmetric stable law with a finite m.g.f. on the hole real line. For $\alpha \in (1, 2)$, the m.g.f. exists for positive h only. Another (non-asymmetric) stable laws have not the exponential moment. So, we deal with the stable laws F_α. Note that the scale parameter c is such that $\varphi(h) = \exp\{h^\alpha/\alpha\}$ for $h > 0$.

The results of the previous sections give us the asymptotic for tails of F_α. Let $F_\alpha(x)$ denote the d.f. of the stable law F_α. By Lemma 2.1, we have

$$\log(1 - F_\alpha(x_n)) \sim -\lambda x_n^{1/\lambda}$$

for every sequence $\{x_n\}$ such that $x_n \to \infty$, where $\lambda = (\alpha - 1)/\alpha$. For $\alpha = 2$, the latter easily follows from the well known relation

$$1 - \Phi(x_n) \sim \frac{1}{\sqrt{2\pi}x_n} e^{-x_n^2/2}.$$

Remember some definitions and facts related to our topic.

We will use the following notation: $g(x) \in SV_a$ ($g(x) \in RV_a$) if $g(x)$ is a slowly (regularly) varying function at a.

The distribution F (and its d.f.) belongs to the domain of attraction of F_α, if there exists a sequence $\{B_n\}$ of positive constants such that the distributions of S_n/B_n converge weakly to F_α where S_n is a sum of n i.i.d. random variables with the distribution F. Note that we always deal with centering at zero.

We write $F \in DN(\alpha)$ or $F \in D(\alpha)$ if the d.f. $F(x)$ belongs to the domain of normal or non-normal attraction of the stable law F_α, correspondingly. The relation $F \in DN(\alpha)$ means that $B_n = bn^{1/\alpha}$ for all $n \in \mathbf{N}$. We always assume that $b = 1$. The relation $F \in D(\alpha)$ means that $B_n = n^{1/\alpha}h(n)$ for all $n \in \mathbf{N}$, where $h(x) \in SV_\infty$ and either $h(x) \to \infty$, or $h(x) \to 0$ as $x \to \infty$.

One usually uses the notation $D(\alpha)$ for the full domain of attraction (i.e. for $DN(\alpha) \cup D(\alpha)$ in our notation), but we prefer to split the domain into two disjoint parts. Moreover, one usually deals with normings $bn^{1/\alpha}$ with $b > 0$. We always assume $b = 1$.

We finish this section with necessary and sufficient conditions for $F \in D(\alpha)$ and $F \in DN(\alpha)$.

Theorem 2.1. *Assume that $EX = 0$ and $\alpha \in (1, 2)$. The following assertions hold:*

1) $F \in DN(2)$ iff $EX^2 = 1$.

2) Assume that $E(X^+)^2 < \infty$. Then $F \in D(2)$ iff $G(x) = \int_{-x}^{0} u^2 dF(u) \in SV_\infty$ and $G(x) \to \infty$ as $x \to \infty$.

3) $F \in DN(\alpha)$ iff $x^\alpha F(-x) \to (\alpha - 1)/(\alpha\Gamma(2-\alpha))$ as $x \to \infty$.

4) $F \in D(\alpha)$ iff $G(x) = x^\alpha F(-x) \in SV_\infty$ and $G(x) \to 0$ or ∞ as $x \to \infty$.

For 2) and 4), the constants B_n may be found from $nB_n^{-2}G(B_n) \to 1$.

Theorem 2.1 follows from results in [Ibragimov and Linnik (1971)]. One has to take into account that we impose additional assumptions with respect to general results on domains of attraction of stable laws. We deal with the norming sequences $n^{1/\alpha}$ for the normal attraction. We assume $E(X^+)^2 < \infty$ for $F \in D(2)$. Moreover, the asymmetry of stable laws yields that right-hand tails are negligible in cases 3) and 4). This implies the difference.

2.4 Functions of Large Deviations Theory and a Classification of Probability Distributions

In this section, we describe the properties of the m.g.f., its logarithmic derivatives and the large deviation function. We also give a classifications of distributions based on these properties. Examples of distributions from each classes are given as well.

Let X be a non-degenerate random variable such that $EX \geq 0$ and $h_0 > 0$. Denote $F(x) = P(X < x)$.

We established in previous sections that the m.g.f. $\varphi(h) = Ee^{hX}$, $0 \leq h < h_0$, has the following properties: $\varphi(0) = 1$, $\varphi(h)$ is increasing and convex, $\log \varphi(h)$ has derivatives of all orders in the interval $0 < h < h_0$. It has also been shown that $m(h) = (\log \varphi(h))'$ is continuous and increasing, $m(0) = EX$, and $\sigma^2(h) = (\log \varphi(h))''$ is continuous, $\sigma^2(0) = DX$ (if the variation exists), $\sigma^2(h) > 0$. It follows that the function $f(h) = hm(h) - \log \varphi(h)$ is continuous and increasing and $f(0) = 0$. Put

$$A = \lim_{h \nearrow h_0} m(h), \quad c_0 = \frac{1}{\lim\limits_{h \nearrow h_0} f(h)}.$$

Here $A \leq \infty$ and $c_0 \geq 0$. We also assume that $1/\infty = 0$ and $1/0 = \infty$. Denote

$$\zeta(x) = f(m^{-1}(x)) \quad \text{for} \quad x \in [EX, A), \tag{2.3}$$

$$\gamma(x) = m(f^{-1}(x)) \quad \text{for} \quad x \in \left[0, \frac{1}{c_0}\right), \tag{2.4}$$

where $m^{-1}(\cdot), f^{-1}(\cdot)$ are the inverse functions to $m(h)$ and $f(h)$, correspondingly.

The function $\zeta(x)$ is called the large deviations function. It plays a key role in what follows.

We see that $\zeta(x)$ is continuous and increasing and $\zeta(EX) = 0$. Moreover,

$$\zeta'(x) = (xm^{-1}(x) - \log \varphi(m^{-1}(x)))' = m^{-1}(x),$$
$$\zeta''(x) = (m^{-1}(x))' > 0.$$

It yields that $\zeta(x)$ is convex and $\gamma(x) = \zeta^{-1}(x)$ is continuous, increasing, concave and $\gamma(0) = EX$.

In what follows, we assume that $\zeta(x)$ and $\gamma(x)$ are defined by relations

$$\zeta(x) = \sup_{h \geq 0:\, \varphi(h) < \infty} \{xh - \log \varphi(h)\},$$
$$\gamma(x) = \sup\{u \geq EX : \zeta(u) \leq x\}.$$

This enlarges domains of definitions and ranges of these functions in some cases.

In the sequel, $\varphi(h)$, $m(h)$, $\sigma^2(h)$, $f(h)$, $\zeta(x)$ and $\gamma(x)$ are called the functions of the LDT.

We will classify distributions basing on values of quantities h_0 and A.

1) Assume first that $h_0 = \infty$ and $A < \infty$. Check that this is only possible for random variables bounded from above. Put $\omega = \operatorname{ess\,sup} X$.

Lemma 2.3. *One has $h_0 = \infty$ and $A < \infty$ if and only if $\omega < \infty$. Moreover, $A = \omega$ in this case.*

Proof. If $A < \infty$, then the definitions of $m(h)$ and A give

$$\log \varphi(h) = \int_0^h m(u)du \le Ah$$

for all $h > 0$. Suppose that $\omega = \infty$. Then $P(X \ge 2A) > 0$ and

$$1 \ge e^{-Ah}\varphi(h) = e^{-Ah}\int_{-\infty}^{\infty} e^{hu}dF(u) \ge e^{-Ah}\int_{2A}^{\infty} e^{hu}dF(u) \ge e^{Ah}P(X \ge 2A) \to \infty$$

as $h \to \infty$. This contradiction shows that $\omega < \infty$.

If $\omega < \infty$, then evidently $h_0 = \infty$. Further, by the definition of ω, we have

$$m(h) = \frac{1}{\varphi(h)} EXe^{hX} \le \omega.$$

It follows that $A \le \omega < \infty$.

To finish the proof, we need only check that $A \ge \omega$. Assume that $A < \omega$. Then there exists $\varepsilon > 0$ such that $A(1+\varepsilon) < \omega$. We have $P(X \ge (1+\varepsilon)A) > 0$ and

$$1 \ge e^{-Ah}\varphi(h) \ge e^{\varepsilon Ah}P(X \ge (1+\varepsilon)A) \to \infty \quad \text{as} \quad h \to \infty.$$

This contradiction yields that $A \ge \omega$. □

a) Suppose that $P(X = \omega) > 0$. By the Lebesgue dominated convergence theorem, we get

$$\varphi(h) = \int_{-\infty}^{\omega} e^{hu}dF(u) = e^{h\omega}\int_{-\infty}^{\omega} e^{h(u-\omega)}dF(u) = e^{h\omega}P(X=\omega)(1+o(1))$$

as $h \to \infty$. Applying again the dominated convergence theorem, we obtain

$$h(m(h) - \omega) = \frac{h}{\varphi(h)}(\varphi'(h) - \omega\varphi(h)) = \frac{e^{h\omega}}{\varphi(h)} \int_{-\infty}^{\omega} h(u - \omega)e^{h(u-\omega)}dF(u) \to 0$$

as $h \to \infty$. Therefore,

$$f(h) = hm(h) - h\omega - \log P(X = \omega) + o(1) \to -\log P(X = \omega)$$

as $h \to \infty$. Hence, $c_0 = -1/\log P(X = \omega) > 0$.

We get additional properties of $\zeta(x)$ and $\gamma(x)$ in this case. We have $\zeta(A) = 1/c_0$ and $\zeta(x) = \infty$ for $x > A$. Indeed, $x > A$ yields

$$xh - \log \varphi(h) \geq xh - Ah \to \infty$$

as $h \to \infty$. Further, $\gamma(x) = \omega$ for $x > 1/c_0$.

If $P(X = \pm 1) = 1/2$, then $h_0 = \infty$, $A = 1$ and $c_0 = 1/\log 2 > 0$.

b) Assume that $P(X = \omega) = 0$. Suppose without loss of generality that $EX > 0$. For $\varepsilon \in (0, c\omega)$, put

$$X_\varepsilon = XI_{\{X<\omega-\varepsilon\}} + \omega I_{\{X\geq\omega-\varepsilon\}}$$

and $\varphi_\varepsilon(h) = Ee^{hX_\varepsilon}$. Here $c \in (0, 1)$ is chosen such that $EX_\varepsilon \geq 0$. In view of $X_\varepsilon \geq X$, we see that

$$xh - \log \varphi_\varepsilon(h) \leq xh - \log \varphi(h)$$

for all $h \geq 0$. It follows that $\zeta_\varepsilon(x) \leq \zeta(x)$ for all $x > EX$, where $\zeta_\varepsilon(x)$ is the large deviation function for X_ε. Then

$$\liminf_{x \nearrow A} \zeta(x) \geq \zeta_\varepsilon(A) = -\log P(X \geq \omega - \varepsilon).$$

Passing to the limit as $\varepsilon \to 0$, we get $\zeta(x) \to \infty$ as $x \to A$. This yields that $c_0 = 0$. In this case, $\gamma(x) \to \omega$ for $x \to \infty$.

If X has the uniform distribution on $[-1, 1]$, then $h_0 = \infty$, $A = 1$ and $c_0 = 0$.

We further assume that at least one of the conditions $h_0 = \infty$ and $A < \infty$ fails. Then $\omega = \infty$ by Lemma 2.3.

2) Suppose that $h_0 = \infty$ and $A = \infty$. We have

$$f(h) = \int_0^h (m(h) - m(u))du \geq \int_0^1 (m(h) - m(u))du \geq m(h) - m(1)$$

for all $h \geq 1$. It follows that $f(h) \to \infty$ as $h \to \infty$ and $c_0 = 0$. Hence $\zeta(x) \to \infty$ and $\gamma(x) \to \infty$ as $x \to \infty$.

If X has the stable distribution F_α then $h_0 = \infty$, $A = \infty$ and $c_0 = 0$. Remember that $\alpha = 2$ for the standard normal law.

3) Assume that $h_0 < \infty$ and $A = \infty$. We prove in the same way as in 2) that

$$f(h) \geq \frac{h_0}{2}\left(m(h) - m\left(\frac{h_0}{2}\right)\right)$$

for all $h \geq h_0/2$. This implies that $f(h) \to \infty$ as $h \to h_0$ and $c_0 = 0$. Therefore, $\zeta(x) \to \infty$ and $\gamma(x) \to \infty$ as $x \to \infty$.

If X has the exponential distribution with the density $p(x) = \mathrm{e}^{-x} I_{(0,\infty)}(x)$, then $h_0 = 1$, $A = \infty$ and $c_0 = 0$.

4) Suppose that $h_0 < \infty$ and $A < \infty$. We have

$$\log \varphi(h_0) = \lim_{h \nearrow h_0} \int_0^h m(u)du \leq \lim_{h \nearrow h_0} Ah = Ah_0 < \infty.$$

Then $f(h) \to h_0 A - \log \varphi(h_0) < \infty$ as $h \to h_0$ and $c_0 = 1/(h_0 A - \log \varphi(h_0)) > 0$.

Note that the functions $\zeta(x)$ and $\gamma(x)$ are defined for large x as well. Indeed, if $x > A$, then the function $xh - \log \varphi(h)$ is increasing in h. Hence

$$\zeta(x) = xh_0 - \log \varphi(h_0)$$

for $x > A$ and

$$\gamma(x) = \frac{x + \log \varphi(h_0)}{h_0}$$

for $x > 1/c_0$. In this case, one can define $f(h)$ and $m(h)$ for $h > h_0$ by formulae

$$m(h) = A + h - h_0, \quad f(h) = \frac{1}{c_0} + h_0(h - h_0). \tag{2.5}$$

Then $\zeta(x) = f(m^{-1}(x))$ for $x \geq EX$ and $\gamma(x) = m(f^{-1}(x))$ for $x > 0$.

If X has the density $p(x) = cx^{-3}\mathrm{e}^{-x} I_{(1,\infty)}(x)$, then $h_0 = 1$, $A = 2$ and $c_0 = 1/(2 - \log(c/2))$.

Thus, there exist five classes of distributions corresponding to various combinations of h_0, A and c_0. Denote these classes as follows:

$$K_1 = \{F(x) : \ h_0 = \infty, \ A < \infty, \ c_0 = 0\},$$
$$K_2 = \{F(x) : \ h_0 = \infty, \ A < \infty, \ c_0 > 0\},$$
$$K_3 = \{F(x) : \ h_0 = \infty, \ A = \infty\},$$
$$K_4 = \{F(x) : \ h_0 < \infty, \ A = \infty\},$$
$$K_5 = \{F(x) : \ h_0 < \infty, \ A < \infty\}.$$

Note that K_1 and K_2 have been considered in items 1,b) and 1,a), correspondingly.

We have also proved that $c_0 = 0$ for $F \in K_3 \cup K_4$ and $c_0 = 1/(h_0 A - \log \varphi(h_0) > 0$ for $F \in K_5$. Moreover, X is bounded from above for $F \in K_1 \cup K_2$ by $\omega = A$ and $c_0 = -1/\log P(X = A)$ for $F \in K_2$.

2.5 Large Deviations and a Non-Invariance

We consider the asymptotic behaviour of $P(S_n \geq bn)$ in this section under $h_0 > 0$. It turns out that the logarithmic asymptotics of these probabilities depend on the large deviations function $\zeta(x)$. This function depends on the full distribution of X and, sometimes, determines this distribution uniquely. This property is called the non-invariance in the sequel.

The main result is as follows.

Theorem 2.2. *Let $X, X_1, X_2, \ldots$ be a sequence of i.i.d. random variables such that $EX \geq 0$ and $h_0 > 0$. Put $S_n = X_1 + X_2 + \cdots + X_n$. Then for all $x \in [EX, A)$, the following relation holds*

$$\log P(S_n \geq xn) \sim -\zeta(x)n. \tag{2.6}$$

Proof. Let h be a solution of the equation $m(h) = x$. This solution exists and it is unique that follows from the properties of $m(h)$ and the definition of A.

By Lemma 2.2 and the definition of $\zeta(x)$, we have

$$\begin{aligned} &\limsup \frac{1}{n} \log P(S_n \geq xn) = \limsup \frac{1}{n} \log P(S_n \geq nm(h)) \\ &\leq -f(h) = -\zeta(x). \end{aligned} \tag{2.7}$$

Take $\varepsilon \in (0,1)$ such that $y = x/(1-\varepsilon) < A$. Let h be a solution of the equation $m(h) = y$. Since h is a fixed number, we have $2\sqrt{n}\sigma(h) = o(nm(h))$ and $2h\sqrt{n}\sigma(h) = o(nf(h))$. By Lemma 2.2, we get

$$\begin{aligned} &P(S_n \geq xn) = P(S_n \geq (1-\varepsilon)yn) \geq P(S_n \geq nm(h) - 2\sqrt{n}\sigma(h)) \\ &\geq \frac{3}{4} e^{-nf(h) - 2h\sqrt{n}\sigma(h)} \geq \frac{3}{4} e^{-nf(h)(1+\varepsilon)} = \frac{3}{4} e^{-n\zeta(y)(1+\varepsilon)} \\ &= \frac{3}{4} e^{-n\zeta(x/(1-\varepsilon))(1+\varepsilon)} \end{aligned}$$

for all sufficiently large n. This yields that

$$\liminf \frac{1}{n} \log P(S_n \geq xn) \geq -\zeta\left(\frac{x}{1-\varepsilon}\right)(1+\varepsilon).$$

Passing to the limit in this inequality as $\varepsilon \to 0$, we obtain

$$\liminf \frac{1}{n} \log P(S_n \geq xn) \geq -\zeta(x).$$

The last inequality and (2.7) imply (2.6). □

Note that the centering of sums S_n in Theorem 2.2 is automatic since the norming and centering (at mean) sequences have the same order n.

If X has c.f. (2.2), then

$$\zeta(x) = \lambda x^{1/\lambda} \quad \text{for} \quad x \geq 0,$$

where $\lambda = (\alpha - 1)/\alpha$. Calculations of $\zeta(x)$ for other distributions show that formulae are much more complicated.

Note that $\zeta(x)$ uniquely determines the distribution of X provided $\varphi(h) < \infty$ for $0 \leq |h| < h_1$. Indeed, $\zeta'(x) = m^{-1}(x)$ and $m(h) = (\log \varphi(h))'$. It turns out that $\zeta(x)$ defines the m.g.f. $\varphi(h)$ in a neighbourhood of zero. At the same time, the c.f. of X is analytic and may be expanded to Taylor's series. The coefficients of this series are uniquely defined by the coefficient of the expansion of $\varphi(h)$. It follows that $\zeta(x)$ defines the distribution of X in this case.

We will show below that for $x = x_n = o(1)$, the behaviour of the probabilities $P(S_n \geq xn)$ depends on the asymptotic of $\zeta(x)$ at 0 provided X is from the domain of attraction of the stable laws with c.f. (2.2).

2.6 Methods of Conjugate Distributions and Truncations

The method of conjugate distributions can not be directly used for random variables without exponential moments. A general way is to present the sum of independent random variables as the sum of truncated from above random variables and a remainder. The method of conjugate distributions may be applied to the sum of truncated random variables and the remainder will be negligible. We develop this techniques below. The aim of this section is to derive an analogue of Lemma 2.2.

Let $X, X_1, X_2, \ldots, X_k$ be non-degenerate, i.i.d. random variables such that $EX = 0$. Fix $y > 0$. Let us introduce random variables

$$\hat{X} = \min\{X, y\} = XI_{\{X<y\}} + yI_{\{X\geq y\}}, \quad \hat{X}_i = \min\{X_i, y\}, \quad i = 1, 2, \ldots, k.$$

Put

$$S_k = \sum_{i=1}^{k} X_i, \quad V_k = \sum_{i=1}^{k} \hat{X}_i - kE\hat{X}.$$

Note that V_k is a centered sum of i.i.d. random variables with exponential moments. It follows that bounds for probability $P(V_k \geq kx)$ may be taken from Lemma 2.2.

For $h \geq 0$ and $x > 0$, put

$$\hat{\varphi}(h) = Ee^{h(\hat{X}-E\hat{X})}, \quad \hat{m}(h) = (\log \hat{\varphi}(h))', \quad \hat{\sigma}^2(h) = (\log \hat{\varphi}(h))'',$$
$$\hat{f}(h) = h\hat{m}(h) - \log \hat{\varphi}(h), \quad \hat{\zeta}(x) = \sup_{h \geq 0}\{xh - \log \hat{\varphi}(h)\}.$$

The next lemma contains an upper bound for $P(S_k \geq kx)$.

Lemma 2.4. *Assume that* $x > 0$, $\rho \in [0,1)$, $\delta \in (0,1)$ *and* $kP(X \geq y) \leq \log 2$. *Then*

$$P(S_k \geq kx) \leq e^{-k\hat{\zeta}(x)} + 2kP(X \geq y).$$

If, in addition, the inequality

$$kP(X \geq y)e^{-k\hat{\zeta}(\delta x)} + (kP(X \geq y))^2 \leq H_1 e^{-k\hat{\zeta}(x)(1-\rho)} \tag{2.8}$$

holds, then

$$P(S_k \geq kx) \leq (1 + H_1)e^{-k\hat{\zeta}(x)(1-\rho)} + kP(X \geq (1-\delta)kx).$$

Proof. Assume first that (2.8) holds. Consider the events $A = \{S_k \geq kx\}$, $A_i = \{X_i = \hat{X}_i\}$ for $i = 1, 2, \ldots, k$ and $B_i = \{\text{exactly } k - i \text{ events from } A_1, A_2, \ldots, A_k \text{ occur}\}$ for $i = 0, 1, 2, \ldots, k$. We have

$$\begin{aligned}
P(S_k \geq kx) &= \sum_{i=0}^{k} P(AB_i) \\
&= P(AA_1 \ldots A_k) + \sum_{i=1}^{k} C_k^i P(AA_1 \ldots A_{k-i}\overline{A_{k-i+1}} \ldots \overline{A_k}) \\
&\leq P(AA_1 \ldots A_k) + kP(AA_1 \ldots A_{k-1}\overline{A_k}) + \sum_{i=2}^{k} C_k^i (P(\overline{A_1}))^i \\
&= a_1 + ka_2 + a_3.
\end{aligned}$$

We derive bounds for a_1, a_2 and a_3 successively. By Lemma 2.2, we get

$$a_1 \leq P\left(\sum_{i=1}^{k} \hat{X}_i \geq kx\right) \leq P(V_k \geq kx) \leq e^{-k\hat{\zeta}(x)}. \tag{2.9}$$

For a_2, we have

$$\begin{aligned}
a_2 &\leq P\left(\sum_{i=1}^{k-1} \hat{X}_i + X_k \geq kx, X_k \geq y\right) \leq P(V_{k-1} + X_k \geq kx, X_k \geq y) \\
&\leq P(V_{k-1} + X_k \geq kx, X_k \geq y, X_k < (1-\delta)kx) + P(X \geq (1-\delta)kx) \\
&\leq P(V_{k-1} \geq \delta kx)P(X \geq y) + P(X \geq (1-\delta)kx).
\end{aligned}$$

The last inequality in (2.9) and the convexity of $\hat{\zeta}(x)$ yield

$$P(V_{k-1} \geq \delta kx) \leq \mathrm{e}^{-(k-1)\hat{\zeta}(\delta kx/(k-1))} \leq \mathrm{e}^{-k\hat{\zeta}(\delta x)}.$$

Hence,

$$a_2 \leq \mathrm{e}^{-k\hat{\zeta}(\delta x)} P(X \geq y) + P(X \geq (1-\delta)kx).$$

Further, making use of condition $kP(X \geq y) \leq \log 2$, we get

$$a_3 \leq \frac{(kP(\overline{A_1}))^2}{2} \sum_{i=2}^{k} \frac{(kP(\overline{A_1}))^{i-2}}{(i-2)!} \leq \frac{(kP(\overline{A_1}))^2}{2} \mathrm{e}^{kP(\overline{A_1})} \leq (kP(X \geq y))^2.$$

Using inequality (2.8), we obtain

$$\begin{aligned} P(S_k \geq kx) &\leq \mathrm{e}^{-k\hat{\zeta}(x)} + kP(X \geq y)\mathrm{e}^{-k\hat{\zeta}(\delta x)} + kP(X \geq (1-\delta)kx) \\ &\qquad + (kP(X \geq y))^2 \\ &\leq (1+H_1)\mathrm{e}^{-k\hat{\zeta}(x)(1-\rho)} + kP(X \geq (1-\delta)kx). \end{aligned}$$

If we do not assume (2.8), then it is clear that $a_2 \leq P(X \geq y)$ and $a_3 \leq kP(x \geq y)$. This yields the result in this case. □

Turn to lower bounds.

Lemma 2.5. *Assume that $x = \hat{m}(h)$ and $\tau \in (0,1)$. If $-E\hat{X} \leq \tau(1-\tau)x$ and $2h\hat{\sigma}(h) \leq \tau\sqrt{k}\hat{f}(h)$, then*

$$P(S_k \geq (1-\tau)^2 kx) \geq \frac{3}{4}\mathrm{e}^{-k\hat{\zeta}(x)(1+\tau)}.$$

Proof. Since $\hat{\varphi}(h) \geq 1$ for all $h > 0$, we get $\hat{f}(h) \leq h\hat{m}(h)$ by the definition of $\hat{f}(h)$. It follows that $2\hat{\sigma}(h) \leq \tau\sqrt{k}\hat{m}(h)$. Making use of Lemma 2.2, we have

$$\begin{aligned} P(S_k \geq (1-\tau)^2 kx) &\geq P\left(\sum_{i=1}^{k} \hat{X}_i \geq (1-\tau)^2 kx\right) \\ &= P(V_k \geq (1-\tau)^2 kx - kE\hat{X})) \geq P(V_k \geq (1-\tau)kx) \\ &\geq P(V_k \geq k\hat{m}(h) - 2\sqrt{k}\hat{\sigma}(h)) \geq \frac{3}{4}\mathrm{e}^{-k\hat{f}(h) - 2h\sqrt{k}\hat{\sigma}(h)} \\ &\geq \frac{3}{4}\mathrm{e}^{-k\hat{f}(h)(1+\tau)} = \frac{3}{4}\mathrm{e}^{-k\hat{\zeta}(x)(1+\tau)}. \end{aligned}$$

The proof is completed. □

Applying Lemma 2.5, we have bounds for $P(S_k \geq kx)$ provided x is from the set of values of $\hat{m}(h)$. We can use the next lemma otherwise.

Lemma 2.6. *Take $k \geq 1$, $\delta > 0$, $x > 0$. If*

$$P(S_{k-1} \geq -\delta kx) - (k-1)P(X \geq (1+\delta)kx) \geq q,$$

then

$$P(S_k \geq kx) \geq qkP(X \geq (1+\delta)kx).$$

Proof. We only need to check that for all u and v, the inequality

$$P(S_k \geq u) \geq kP(X \geq u+v)\,(P(S_{k-1} \geq -v) - (k-1)P(X \geq u+v)) \quad (2.10)$$

holds.

Put $A_j = \{X_j \geq u+v\}$, $B_j = \{S_k - X_j \geq -v\}$, $j = 1, 2, \ldots, k$. We have

$$P(S_k \geq u) \geq P(\cup_{j=1}^k A_j B_j) \geq \sum_{j=1}^k \Big(P(A_j B_j) - \sum_{\substack{n=1\\ n\neq j}}^k P(A_j A_n B_j B_n)\Big)$$

$$\geq \sum_{j=1}^k \Big(P(A_j)P(B_j) - \sum_{\substack{n=1\\ n\neq j}}^k P(A_j A_n)\Big) = \sum_{j=1}^k P(A_j)\Big(P(B_j) - \sum_{\substack{n=1\\ n\neq j}}^k P(A_n)\Big).$$

The latter coincides with the right-hand side of (2.10). □

2.7 Asymptotic Expansions of Functions of Large Deviations Theory in Case of Finite Variations

Asymptotic expansions for the functions of the LDT at zero play a key role in proofs below. In this section, we deal with the case of finite variations.

Our first result is for random variables with exponential moments.

Lemma 2.7. *If $EX = 0$, $EX^2 = 1$ and $h_0 > 0$, then*

$$\varphi(h) = 1 + \frac{h^2}{2}(1 + o(1)), \quad m(h) = h(1 + o(1)), \quad \sigma^2(h) = 1 + o(1),$$

$$f(h) = \frac{h^2}{2}(1 + o(1)) \quad \text{as} \quad h \to 0.$$

Proof. The conditions $EX = 0$ and $EX^2 = 1$, the definitions of $m(h)$ and $\sigma^2(h)$ and their properties imply that $\varphi'(0) = m(0) = 0$ and $\varphi''(0) = \sigma^2(0) = 1$.

Take $h \in (0, h_0)$. By Taylor's formula, we have

$$\varphi(h) = \varphi(0) + h\varphi'(0) + \frac{h^2}{2}\varphi''(\theta) = 1 + \frac{h^2}{2}\varphi''(\theta),$$

where $\theta \in (0, h)$. Hence, $\varphi''(\theta) \to 1$ and

$$\varphi(h) = 1 + \frac{h^2}{2}(1 + o(1))$$

as $h \to 0$. Further,

$$m(h) = \frac{\varphi'(h)}{\varphi(h)} = \frac{\varphi'(0) + h\varphi''(\theta_1)}{1 + o(1)} = \frac{h(1 + o(1))}{1 + o(1)} = h(1 + o(1))$$

as $h \to 0$. The continuity of $\sigma^2(h)$ yields $\sigma^2(h) = 1 + o(1)$ as $h \to 0$. It remains to apply the definition of $f(h)$. □

Note that the asymptotics in Lemma 2.7 coincide with those for the standard normal random variable.

Lemmas 2.7 and 2.2 allow to obtain large deviations results for $h_0 > 0$.

We now turn to random variables without exponential moments. Using truncations, we derived absolute inequalities for $P(S_k \geq kx)$ in the last section. In the next section, they will be applied with $k = n$, $x = x_n$ and $y = y_n \to \infty$ to study large deviations. To this end, we need asymptotic expansions for the functions of LDT for $\hat{X} - E\hat{X}$ with $y = y_n$. In this section, we find conditions under which these functions have the same asymptotics as in Lemma 2.7.

Let X be a random variable with a d.f. $F(x)$. Let $\{y_n\}$ be a sequence of positive numbers such that $y_n \to \infty$. For $n \in \mathbf{N}$, put

$$Y_n = \min\{X, y_n\} \quad Z_n = Y_n - EY_n, \quad F_n(x) = P(Z_n < x).$$

For $h \geq 0$ and $x > 0$, define functions

$$\varphi_n(h) = Ee^{hZ_n}, \quad m_n(h) = (\log \varphi_n(h))', \quad \sigma_n^2(h) = (\log \varphi_n(h))'',$$
$$f_n(h) = hm_n(h) - \log \varphi_n(h), \quad \zeta_n(x) = \sup_{h \geq 0}\{xh - \log \varphi_n(h)\},$$

where $n \in \mathbf{N}$. It is clear that for every fixed n, these functions coincide correspondingly with $\hat{\varphi}(h)$, $\hat{m}(h)$, $\hat{f}(h)$ and $\hat{\zeta}(x)$ when $y = y_n$. So, they are the functions of the LDT for the random variable $\hat{X} - E\hat{X}$ with $y = y_n$.

It is convenient to put $y_n = \infty$ for all n when $h_0 > 0$ and we need no truncations. Of course, we then assume that $h < h_0$ and $x < A$ in the above formulae.

Our next result is as follows.

Lemma 2.8. *Assume that* $EX = 0$, $EX^2 = 1$, $h = h_n \to 0$ *and one of the following conditions holds:*

1) $h_0 > 0$ *and* $y = y_n = \infty$ *for all* n,

2) $y = y_n \to \infty$ *and*

$$h \int_0^\infty x^3 e^{hx} dF_n(x) = o(1). \tag{2.11}$$

Then

$$\varphi_n(h) = 1 + \frac{h^2}{2}(1 + o(1)), \tag{2.12}$$

$$m_n(h) = h(1 + o(1)), \tag{2.13}$$

$$\sigma_n^2(h) = 1 + o(1), \tag{2.14}$$

$$f_n(h) = \frac{h^2}{2}(1 + o(1)). \tag{2.15}$$

Proof. If condition 1) holds, then Lemma 2.7 yields (2.12)–(2.15). Assume that condition 2) holds.

For $k = 0, 1, 2$, denote

$$e_k(x) = e^x - \sum_{j=0}^{k} \frac{x^j}{j!},$$

$$L_{k,n} = \int_{-\infty}^{0} x^k e_{2-k}(xh) dF_n(x), \quad I_{k,n} = \int_0^{+\infty} x^k e_{2-k}(xh) dF_n(x).$$

Then

$$\varphi_n(h) = \int_{-\infty}^{\infty} e^{hx} dF_n(x) = 1 + \frac{h^2 E Z_n^2}{2} + L_{0,n} + I_{0,n}, \tag{2.16}$$

$$\varphi_n(h) m_n(h) = \int_{-\infty}^{\infty} x e^{hx} dF_n(x) = h E Z_n^2 + L_{1,n} + I_{1,n}, \tag{2.17}$$

$$(\sigma_n^2(h) + (m_n(h))^2)\varphi_n(h) = \int_{-\infty}^{\infty} x^2 e^{hx} dF_n(x) = E Z_n^2 + L_{2,n} + I_{2,n}. \tag{2.18}$$

We will use the following well known result.

Lemma 2.9. *For all* $x > 0$, *the inequalities*

$$e_m(x) \le \frac{x^{m+1}}{(m+1)!} e^x, \quad C \frac{x^{m+1}}{1+x} \le |e_m(-x)| \le D \frac{x^{m+1}}{1+x}, \quad m = 0, 1, 2,$$

hold, where C *and* D *are absolute positive constants.*

By Lemma 2.9 with $m = 2 - k$ and (2.11), we have

$$I_{k,n} \leq \frac{h^{3-k}}{(3-k)!} \int_0^{+\infty} x^3 e^{hx} dF_n(x) = o(h^{2-k}), \quad k = 0, 1, 2. \tag{2.19}$$

Applying again Lemma 2.9 with $m = 2 - k$, we obtain

$$Ch^{2-k} I(n) \leq |L_{k,n}| \leq Dh^{2-k} I(n), \quad k = 0, 1, 2,$$

where

$$I(n) = \int_{-\infty}^{0} \frac{h|x|^3}{1 + |hx|} dF_n(x).$$

Denote $v_n(u) = \int_{-\infty}^{u} x^2 dF_n(x)$. Then

$$I(n) \leq \int_{-\infty}^{-1/h} x^2 dF_n(x) + h \int_{-1/h}^{0} |x|^3 dF_n(x) = v_n\left(-\frac{1}{h}\right) + h \int_{-1/h}^{0} |x| dv_n(x)$$

$$= h\Big(\int_0^{1/\sqrt{h}} + \int_{1/\sqrt{h}}^{1/h} \Big) v_n(-x) dx \leq \sqrt{h} v_n(0) + (1 - \sqrt{h}) v_n\left(-\frac{1}{\sqrt{h}}\right).$$

Since

$$v_n(u) = \int_{-\infty}^{u+EY_n} x^2 dF(x) - 2EY_n \int_{-\infty}^{u+EY_n} x dF(x) + (EY_n)^2 \int_{-\infty}^{u+EY_n} dF(x)$$

and $EY_n \to 0$, we have $v_n(0) \to E(X^-)^2$ and $v_n(-1/\sqrt{h}) \to 0$. It follows that $I(n) = o(1)$ and

$$L_{k,n} = o(h^{2-k}), \quad k = 0, 1, 2. \tag{2.20}$$

It is clear that $EY_n \to 0$, $EY_n^2 \to 1$ and $EZ_n^2 = EY_n^2 - (EY_n)^2 \to 1$.

Relations (2.16), (2.19) and (2.20) imply (2.12). Assertions (2.12), (2.17), (2.19) and (2.20) yield (2.13). From (2.12), (2.13), (2.18), (2.19) and (2.20), we obtain (2.14). The definition of the function $f_n(h)$, relations (2.12) and (2.13) imply (2.15). □

In the next lemma, we replace condition (2.11) on distributions of truncated random variables by those on the distributions of summands.

Lemma 2.10. *Assume that* $EX = 0$, $EX^2 = 1$, $y = y_n \to \infty$, $h = h_n \to 0$ *and one of the following conditions holds:*

1) there exists $\beta \in (0,1)$ *such that* $\limsup h_n y_n^{1-\beta} < 1$ *and* $\mathrm{E}e^{(X^+)^\beta} < \infty$,

2) $h_n y_n = O(1)$.

Then relations (2.12)–(2.15) hold.

Proof. To prove the lemma, we check condition (2.11) and apply Lemma 2.8.

Denote $z_n = \operatorname{ess\,sup}(Z_n)$. Note that $z_n = y_n - EY_n = y_n + o(1)$. We have

$$J_n = \int_0^\infty x^3 e^{hx} dF_n(x) = \int_0^{z_n} x^3 e^{hx} dF(x + EY_n) + z_n^3 e^{hz_n} P(X \geq y).$$

Further, we get

$$I_n = \int_0^{z_n} x^3 e^{hx} dF(x + EY_n) = \int_{EY_n}^{y_n} (x - EY_n)^3 e^{h(x-EY_n)} dF(x)$$

$$= \left(\int_{EY_n}^{-EY_n} + \int_{-EY_n}^{y_n} \right) (x - EY_n)^3 e^{h(x-EY_n)} dF(x)$$

$$\leq 8|EY_n|^3 e^{-2hEY_n} + 8e^{-hEY_n} \int_{-EY_n}^{y_n} x^3 e^{hx} dF(x) \leq C_0 \int_0^{y_n} x^3 e^{hx} dF(x).$$

Assume first that $hy = O(1)$. Then

$$J_n \leq C_1 \int_0^{y_n} x^3 dF(x) + C_1 z_n^3 P(X \geq y).$$

In view of $EX^2 < \infty$, there exists a positive, even function $g(x)$ such that $g(x) \to \infty$ as $x \to \infty$, $g(x)$ and $x/g(x)$ are non-decreasing for $x > 0$ and $EX^2 g(X) < \infty$. Hence,

$$\int_0^{y_n} x^3 dF(x) = \int_0^{y_n} x^2 g(x) \frac{x}{g(x)} dF(x) \leq C_2 \frac{y_n}{g(y_n)},$$

and

$$P(X \geq y) \leq C_2 \frac{1}{y^2 g(y)}.$$

This yields that

$$hJ_n \leq C_3 \frac{hy_n}{g(y_n)} + C_3 \frac{hz_n^3}{y^2 g(y)} \leq C_4 \frac{1}{g(y_n)} \to 0$$

and relation (2.11) holds.

Assume that condition 1) is satisfied. Let a be a positive real number such that $hy^{1-\beta} \le a < 1$ for all sufficiently large n. Then

$$\int_0^{y_n} x^3 \mathrm{e}^{hx} dF(x) \le \int_0^{y_n} x^3 \mathrm{e}^{ax^\beta} dF(x) \le \sup_{x\ge 0} \left\{x^3 \mathrm{e}^{(a-1)x^\beta}\right\} \int_0^{y_n} \mathrm{e}^{x^\beta} dF(x) \le C_5$$

for all sufficiently large n. Further, we have

$$z_n^3 \mathrm{e}^{hz_n} P(X \ge y) \le C_6 y^3 \mathrm{e}^{hy} P(X \ge y) \le C_7 y^3 \mathrm{e}^{hy} \mathrm{e}^{-y^\beta} \le C_7 y^3 \mathrm{e}^{(a-1)y^\beta} \to 0.$$

It follows that $hJ_n = o(1)$ and we get (2.11) again. □

In the next lemma, we deal with a special choice of h.

Lemma 2.11. *Assume that $EX = 0$, $EX^2 = 1$, $y = y_n \to \infty$, $u_n \to 0$ and one of the following conditions holds:*

1) there exists $\beta \in (0,1)$ such that $\limsup u_n y_n^{1-\beta} < 1/2$ and $E\mathrm{e}^{(X^+)^\beta} < \infty$,

2) $u_n y_n = O(1)$.

For every fixed n, let h_n be a solution of the equation $m_n(h) = u_n$.

Then relations (2.12)–(2.15) hold.

Proof. For every fixed n, we have by Lemma 2.3 that $m_n(h) \nearrow y_n - EY_n$ as $h \nearrow \infty$. It follows that the equation $m_n(h) = u_n$ has a unique solution h_n provided n is fixed and large enough.

Lemma 2.10 yields that $m_n(2u_n) = 2u_n(1 + o(1))$. It follows that $m(2u_n) \ge u_n = m_n(h_n)$ for all large n. Since $m_n(h)$ is an increasing function of h, we arrive at $h_n \le 2u_n$ for all large n. Lemma 2.11 now follows from Lemma 2.10. □

2.8 Large Deviations in Case of Finite Variations

In this section, we find the logarithmic asymptotics of large deviation probabilities for sums of i.i.d. random variables with a finite second moment. We first show that if $h_0 > 0$ and $x_n = o(\sqrt{n})$, then this asymptotic coincides with that in the Gaussian case. Further, we deal with the case $h_0 = 0$. We prove that the asymptotic is Gaussian for x_n from regions depending on moment assumptions. The Linnik condition yields the result in a power region while $E(X^+)^p < \infty$ provides a logarithmic zone only.

Assume first that $h_0 > 0$.

Theorem 2.3. *Let $X, X_1, X_2, \ldots$ be a sequence of i.i.d. random variables such that $EX = 0$, $EX^2 = 1$ and $h_0 > 0$. Put $S_n = X_1 + X_2 + \cdots + X_n$. Then for every sequence $\{x_n\}$ with $x_n \to \infty$ and $x_n = o(\sqrt{n})$, the following relation holds*

$$\log P(S_n \geq x_n\sqrt{n}) \sim -\frac{x_n^2}{2}. \tag{2.21}$$

Proof. Let $h = h_n$ be a solution of the equation $m(h) = x_n/\sqrt{n}$. Since $m(h)$ is increasing and $m(0) = 0$, this equation has a unique solution for all sufficiently large n. Moreover, $h \to 0$. Making use of Lemma 2.7, we obtain

$$h = \frac{x_n}{\sqrt{n}}(1 + o(1)), \quad nf(h) = n\frac{h^2}{2}(1 + o(1)) = \frac{x_n^2}{2}(1 + o(1)).$$

Take $\varepsilon \in (0, 1)$. By Lemma 2.2, we have

$$\log P(S_n \geq x_n\sqrt{n}) \leq -\frac{x_n^2}{2}(1 - \varepsilon)$$

for all sufficiently large n. It follows that

$$\limsup \frac{1}{x_n^2} \log P(S_n \geq x_n\sqrt{n}) \leq -\frac{1 - \varepsilon}{2}.$$

Passing in this relation to the limit as $\varepsilon \to 0$, we arrive at

$$\limsup \frac{1}{x_n^2} \log P(S_n \geq x_n\sqrt{n}) \leq -\frac{1}{2}. \tag{2.22}$$

Take $\varepsilon \in (0, 1)$. Put $y_n = x_n/(1 - \varepsilon)$. Let $h = h_n$ be a solution of the equation $m(h) = y_n/\sqrt{n}$. Using Lemma 2.7, we have

$$\begin{aligned} nm(h) - 2\sqrt{n}\sigma(h) &= y_n\sqrt{n}(1 + o(1)) + O(\sqrt{n}) \\ &= y_n\sqrt{n}(1 + o(1)) = \frac{x_n\sqrt{n}(1 + o(1))}{1 - \varepsilon}. \end{aligned}$$

By Lemma 2.2, we get

$$P(S_n \geq x_n\sqrt{n}) \geq P(S_n \geq nm(h) - 2\sqrt{n}\sigma(h)) \geq \frac{3}{4}\mathrm{e}^{-nf(h) - 2h\sqrt{n}\sigma(h)}$$

for all sufficiently large n. Applying of Lemma 2.7 yields

$$nf(h) - 2h\sqrt{n}\sigma(h) = \frac{y_n^2}{2}(1 + o(1)) + O(y_n) = \frac{y_n^2}{2}(1 + o(1)) = \frac{x_n^2(1 + o(1))}{2(1 - \varepsilon)^2}.$$

It follows that

$$\log P(S_n \geq x_n\sqrt{n}) \geq \log\frac{3}{4} - \frac{x_n^2}{2(1 - \varepsilon)^3}$$

for all sufficiently large n. Hence

$$\liminf \frac{1}{x_n^2} \log P(S_n \geq x_n\sqrt{n}) \geq -\frac{1}{2(1-\varepsilon)^3}.$$

Passing in this inequality to the limit as $\varepsilon \to 0$, we get

$$\liminf \frac{1}{x_n^2} \log P(S_n \geq x_n\sqrt{n}) \geq -\frac{1}{2}.$$

This relation and (2.22) imply (2.21). □

We now derive analogues of Theorem 2.3 when the one-sided Cramér condition is violated.

Theorem 2.4. *Let $X, X_1, X_2, \ldots$ be a sequence of i.i.d. random variables such that $EX = 0$, $EX^2 = 1$ and $E\mathrm{e}^{(X^+)^\beta} < \infty$ for some $\beta \in (0,1)$. Then relation (2.21) holds for every sequence $\{x_n\}$ with $x_n \to \infty$ and $x_n = o(n^{\beta/(4-2\beta)})$.*

Remember that $E\mathrm{e}^{|X|^\beta} < \infty$ for some $\beta \in (0,1)$ is the Linnik condition. In Theorem 2.4, we assume the Linnik condition for X^+ instead of $|X|$ which is called the one-sided Linnik condition.

Proof. Put $u_n = x_n/\sqrt{n}$ and $y = y_n = x_n\sqrt{n}$ for $n \in \mathbf{N}$. We have $u_n y^{1-\beta} = x_n^{2-\beta} n^{-\beta/2} = o(1)$ and condition 1) of Lemma 2.11 is fulfilled. Let $h = h_n$ be a solution of the equation $m_n(h) = u_n$. By Lemma 2.11, relations (2.12)–(2.15) hold.

Put $k = n$, $x = m_n(h)$. Note that $x = \frac{x_n}{\sqrt{n}} = h(1+o(1))$ in view of (2.13).

Verify condition (2.8) of Lemma 2.4. By (2.15), we have

$$k\hat{\zeta}(x) = k\hat{f}(h) = nf_n(h) = n\frac{h^2}{2}(1+o(1)) = \frac{x_n^2}{2}(1+o(1)).$$

Using the one-sided Linnik condition, we get

$$kP(X \geq y) = nP(X \geq x_n\sqrt{n}) = o\left(n\mathrm{e}^{-x_n^\beta n^{\beta/2}}\right) = o(1).$$

Hence,

$$\begin{aligned} &kP(X \geq y)\mathrm{e}^{-k\hat{\zeta}(\delta x)} + (kP(X \geq y))^2 \leq 2kP(X \geq y) \\ &= o\left(n\mathrm{e}^{-x_n^\beta n^{\beta/2}}\right) = o\left(\mathrm{e}^{-k\hat{\zeta}(x)}\right), \end{aligned}$$

and condition (2.8) holds with $H_1 = 1$ for all sufficiently large n. It follows by Lemma 2.4 that

$$P(S_k \geq kx) \leq 2\mathrm{e}^{-k\hat{\zeta}(x)} + kP(X \geq (1-\delta)kx)$$

for all sufficiently large n. Taking into account (2.13) and the one-sided Linnik condition, we have

$$kP(X \geq (1-\delta)kx) \leq nP(X \geq (1-\delta)^2 x_n\sqrt{n})$$
$$= o\left(n\mathrm{e}^{-(1-\delta)^{2\beta}x_n^\beta n^{\beta/2})}\right) = o\left(\mathrm{e}^{-k\hat{\zeta}(x)}\right).$$

This implies that

$$P(S_k \geq kx) \leq 3\mathrm{e}^{-k\hat{\zeta}(x)}$$

for all sufficiently large n. Hence, for every $\varepsilon > 0$,

$$P(S_n \geq x_n\sqrt{n}) \leq 3\mathrm{e}^{-(1-\varepsilon)x_n^2/2} \tag{2.23}$$

for all sufficiently large n. It follows that

$$\limsup \frac{1}{x_n^2} \log P(S_n \geq x_n\sqrt{n}) \leq -\frac{1-\varepsilon}{2}.$$

Passing to the limit as $\varepsilon \to 0$, we conclude that

$$\limsup \frac{1}{x_n^2} \log P(S_n \geq x_n\sqrt{n}) \leq -\frac{1}{2}. \tag{2.24}$$

Test the conditions of Lemma 2.5. Take $\delta \in (0,1)$. We have

$$E\hat{X} = -\int_y^\infty u dF(u) + yP(X \geq y) = o\left(y\mathrm{e}^{-y^\beta}\right) = o\left(\mathrm{e}^{-(1-\delta)x_n^\beta n^{\beta/2}}\right).$$

This and assertions (2.14) and (2.15) imply that $-E\hat{X} = o(x)$, $h\sigma_n(h) = o(\sqrt{n}f_n(h))$, and the conditions of Lemma 2.5 are satisfied. Hence, for every $\tau \in (0,1)$, the inequalities

$$P(S_n \geq (1-\tau)^3 x_n\sqrt{n}) \geq P(S_k \geq (1-\tau)^2 kx)$$
$$\geq \mathrm{e}^{-k\hat{\zeta}(x)(1+\tau)} \geq \mathrm{e}^{-x_n^2(1+\tau)^2/2} \tag{2.25}$$

hold for all sufficiently large n. Then

$$\liminf \frac{1}{x_n^2} \log P(S_n \geq (1-\tau)^3 x_n\sqrt{n}) \geq -\frac{(1+\tau)^2}{2}.$$

Making the replacement $x_n = (1-\tau)^3 x_n$ in the last inequality, we have

$$\liminf \frac{1}{x_n^2} \log P(S_n \geq x_n\sqrt{n}) \geq -\frac{(1+\tau)}{2(1-\tau)^6}.$$

Passing to the limit as $\tau \to 0$ in the latter inequality, we obtain

$$\liminf \frac{1}{x_n^2} \log P(S_n \geq x_n\sqrt{n}) \geq -\frac{1}{2}. \tag{2.26}$$

Relation (2.21) follows from (2.24) and (2.26). □

We now replace the one-sided Linnik condition by a weaker power condition.

Theorem 2.5. *Let $X, X_1, X_2, \ldots$ be a sequence of i.i.d. random variables such that $EX = 0$, $EX^2 = 1$ and $E(X^+)^p < \infty$, $p > 2$. Then relation (2.21) holds for every sequence $\{x_n\}$ with $x_n \to \infty$ and* $\limsup x_n^2(\log n)^{-1} \le p-2$.

Proof. We need the next technical lemma.

Lemma 2.12. *If $x_n^2 \le (1+\tau)((p-2)\log n + p\log\log n)$ for $\tau > 0$ and $n \ge n_0$, then*

$$x_n^{-p} n^{(2-p)/2} \le (p-2)^{-p/2} \mathrm{e}^{-x_n^2/(2+2\tau)}$$

for $n \ge n_0$.

Proof. The function $u^{-p}\mathrm{e}^{u^2/2}$ strictly increases for $u > \sqrt{p}$. If $x_n^2 \le (p-2)\log n$, then

$$x_n^{-p}\mathrm{e}^{x_n^2/2} \le (p-2)^{-p/2}(\log n)^{-p/2} n^{(p-2)/2}$$

and, consequently,

$$x_n^{-p} n^{(2-p)/2} \le (p-2)^{-p/2}(\log n)^{-p/2}\mathrm{e}^{-x_n^2/2}.$$

If $x_n^2 \ge (p-2)\log n$, then

$$x_n^{-p} n^{(2-p)/2} \le (p-2)^{-p/2}(\log n)^{-p/2} n^{(2-p)/2} \le (p-2)^{-p/2}\mathrm{e}^{-x_n^2/(2+2\tau)}.$$

The lemma follows. □

Put $u_n = x_n/\sqrt{n}$ and $y = y_n = \sqrt{n}/x_n$ for $n \in \mathbf{N}$. Let $h = h_n$ be a solution of the equation $m_n(h) = u_n$. Since $u_n y = 1$, relations (2.12)–(2.15) hold by Lemma 2.11.

Put $k = n$, $x = m_n(h)$. Note that $x = \frac{x_n}{\sqrt{n}} = h(1+o(1))$ in view of (2.13).

Check condition (2.8) of Lemma 2.4. By (2.15), we have

$$k\hat{\zeta}(\delta x) = \delta^2 \frac{x_n^2}{2}(1+o(1)),$$

where $\delta \in (0,1]$.

Making use of the condition $E(X^+)^p < \infty$ and Lemma 2.12, we get

$$kP(X \ge y) = nP\left(X \ge \frac{\sqrt{n}}{x_n}\right) = o\left(x_n^p n^{(2-p)/2}\right) = o(1).$$

This yields that

$$kP(X \geq y)\mathrm{e}^{-k\hat{\zeta}(\delta x)} + (kP(X \geq y))^2 = o\left(x_n^{2p}\mathrm{e}^{-(1+\delta^2)x_n^2/2}\right) + o\left(x_n^{2p}n^{2-p}\right)$$
$$= o\left(x_n^{2p}\mathrm{e}^{-(1+\delta^2)x_n^2/2}\right) + o\left(x_n^{4p}\mathrm{e}^{-x_n^2}\right) = o\left(\mathrm{e}^{-k\hat{\zeta}(x)}\right),$$

and condition (2.8) is fulfilled with $H_1 = 1$ for all sufficiently large n. It follows by Lemma 2.4 that

$$P(S_k \geq kx) \leq 2\mathrm{e}^{-k\hat{\zeta}(x)} + kP(X \geq (1-\delta)kx)$$

for all sufficiently large n. Taking into account (2.13) and condition $E(X^+)^p < \infty$, we have

$$kP(X \geq (1-\delta)kx) \leq nP(X \geq (1-\delta)^2 x_n\sqrt{n}) = o\left(x_n^{-p}n^{(2-p)/2}\right).$$

Take $\varepsilon > 0$. Put $\tau = \varepsilon/(1-\varepsilon)$. By Lemma 2.12, inequality (2.23) holds for all sufficiently large n. In the same way as in the proof of Theorem 2.4, we get (2.24).

Check now the conditions of Lemma 2.5. We have

$$E\hat{X} = -\int_y^\infty u dF(u) + yP(X \geq y) = o\left(y^{1-p}\right) = o\left(x_n^{p-1}n^{(1-p)/2}\right).$$

This and relations (2.14) and (2.15) yield that $-E\hat{X} = o(x)$, $h\sigma_n(h) = o(\sqrt{n}f_n(h))$, and the conditions of Lemma 2.5 hold.

The remainder repeats the end of the proof of the previous theorem. By Lemma 2.5, for every $\tau \in (0,1)$ inequality (2.25) holds for all sufficiently large n. This yields (2.26).

Relation (2.21) follows from (2.24) and (2.26). □

Theorems 2.3–2.5 imply that weaker moment assumptions give narrower zones in which large deviations have the Gaussian asymptotic. If the one-sided Linnik condition is satisfied, then this zone is $o(n^{\beta/(4-2\beta)})$. For $\beta = 1$, we arrive at the Cramér zone which is widest possible by Theorem 2.2. If $E(X^+)^p < \infty$ for some $p > 2$, then the zone is logarithmic.

Moreover, Theorems 2.3–2.5 yield that the behaviour of $\log P(S_n \geq x_n\sqrt{n})$ depends on the right-hand tail of the distribution of X more than on the left-hand one. The conditions of Theorems 2.3–2.5 are not symmetric in this sense. We only assume $E(X^-)^2 < \infty$. If we impose the same conditions for X^- as for X^+, then we get the Gaussian asymptotics for $\log P(S_n \leq -x_n\sqrt{n})$ as well.

2.9 Asymptotic Expansions of Functions of Large Deviations Theory for $D(2)$

In this section, we derive asymptotic expansions of the functions of the LDT for random variables from the domains of non-normal attraction of the normal law.

We first prove the following analogue of Lemma 2.8. The case of the finite exponential moment is included in there as well.

Lemma 2.13. *Assume that* $EX = 0$, $E(X^+)^2 < \infty$ *and* $F \in D(2)$. *Assume that* $h = h_n \to 0$ *and one of the following conditions holds:*

1) $h_0 > 0$ *and* $y = y_n = \infty$ *for all* n,

2) $y = y_n \to \infty$ *and*

$$\int_0^\infty x^2 e^{hx} dF_n(x) = O(1). \tag{2.27}$$

Then

$$\varphi_n(h) = 1 + \frac{h^2}{2} G\left(\frac{1}{h}\right)(1 + o(1)), \tag{2.28}$$

$$m_n(h) = hG\left(\frac{1}{h}\right)(1 + o(1)), \tag{2.29}$$

$$\sigma_n^2(h) = G\left(\frac{1}{h}\right)(1 + o(1)), \tag{2.30}$$

$$f_n(h) = \frac{h^2}{2} G\left(\frac{1}{h}\right)(1 + o(1)), \tag{2.31}$$

where $G(x) = \int_{-x}^{0} u^2 dF(u)$, $x > 0$.

Note that if condition 1) holds then $\varphi_n(h) = \varphi(h)$, $m_n(h) = m(h)$, $\sigma_n^2(h) = \sigma^2(h)$ and $f_n(h) = f(h)$ for $h \in (0, h_0)$ and all n.

Proof. Assume that condition 2) holds.

Put $e_{-1}(x) = e^x$. For $k = 0, 1, 2$, denote

$$e_k(x) = e^x - \sum_{j=0}^{k} \frac{x^j}{j!},$$

$$L_{k,n} = \int_{-\infty}^{0} x^k e_{1-k}(xh) dF_n(x), \quad I_{k,n} = \int_0^{+\infty} x^k e_{1-k}(xh) dF_n(x).$$

Then we have

$$\varphi_n(h) = \int_{-\infty}^{\infty} e^{hx} dF_n(x) = 1 + L_{0,n} + I_{0,n}, \tag{2.32}$$

$$\varphi_n(h) m_n(h) = \int_{-\infty}^{\infty} x e^{hx} dF_n(x) = L_{1,n} + I_{1,n}, \tag{2.33}$$

$$(\sigma_n^2(h) + (m_n(h))^2)\varphi_n(h) = \int_{-\infty}^{\infty} x^2 e^{hx} dF_n(x) = L_{2,n} + I_{2,n}. \tag{2.34}$$

By Lemma 2.9 and (2.27), we have in the same way as in Lemma 2.8 that

$$I_{k,n} \leq h^{2-k} \int_0^{\infty} x^2 e^{hx} dF_n(x) = O(h^{2-k}), \quad k = 0, 1, 2. \tag{2.35}$$

To estimate $L_{k,n}$, we need the following result on properties of $G(x)$.

Lemma 2.14. *If the conditions of Lemma 2.13 hold then $G(x) \in SV_\infty$ and*

$$x \int_{|u|>x} |u| dF(u) = o(G(x)), \tag{2.36}$$

$$x^2 P(|X| > x) = o(G(x)), \tag{2.37}$$

$$\int_0^x G(u) du \sim x G(x), \tag{2.38}$$

$$\int_{-x}^{0} u^3 dF(u) = o(xG(x)) \tag{2.39}$$

as $x \to \infty$.

Relations (2.36)–(2.38) may be found in [Feller (1971)]. Integrating by parts, one can easily derive relation (2.39) from (2.38).

We write

$$L_{k,n} = \left(\int_{-\infty}^{-1/h} + \int_{-1/h}^{0} \right) x^k e_{1-k}(xh) dF_n(x) = \bar{L}_{k,n} + \hat{L}_{k,n}.$$

Making use of Lemma 2.9, we get

$$|\bar{L}_{k,n}| \le Dh^{-k} \int_{-\infty}^{-1/h} \frac{|xh|^2}{1+|xh|} dF_n(x) \le Dh^{1-k} \int_{-\infty}^{-1/h} |x| dF_n(x)$$

$$= Dh^{1-k} \int_{-\infty}^{-1/h+EY_n} |x - EY_n| dF(x) \le Dh^{1-k} \int_{-\infty}^{-1/h} |x - EY_n| dF(x)$$

$$\le Dh^{1-k} \int_{-\infty}^{-1/h} |x| dF(x) + Dh^{1-k} |EY_n| P\left(|X| \ge \frac{1}{h}\right), \quad k = 0, 1, 2.$$

By (2.36) and (2.37), we arrive at

$$\bar{L}_{k,n} = o\left(h^{2-k} G\left(\frac{1}{h}\right)\right), \quad k = 0, 1, 2. \tag{2.40}$$

Further, for $k = 0, 1, 2$, we write

$$\hat{L}_{k,n} = \int_{-1/h}^{0} x^k e_{2-k}(xh) dF_n(x) + \frac{h^{2-k}}{(2-k)!} \int_{-1/h}^{0} x^2 dF_n(x).$$

We have

$$\int_{-1/h}^{0} x^2 dF_n(x) = \int_{-1/h+EY_n}^{EY_n} (x - EY_n)^2 dF(x) = \int_{-1/h+EY_n}^{EY_n} x^2 dF(x) + o(1)$$

$$= \int_{-1/h+EY_n}^{0} x^2 dF(x) + o(1) = G\left(\frac{1}{h}\right)(1 + o(1)).$$

Moreover, by Lemma 2.9, (2.37) and (2.39), we get

$$\int_{-1/h}^{0} |x^k e_{2-k}(xh)| dF_n(x) \leq Dh^{3-k} \int_{-1/h}^{0} |x|^3 dF_n(x)$$

$$= Dh^{3-k} \int_{-1/h+EY_n}^{EY_n} |x - EY_n|^3 dF(x)$$

$$\leq Dh^{3-k} \int_{-1/h+EY_n}^{0} |x - EY_n|^3 dF(x)$$

$$\leq 8Dh^{3-k} \left(\int_{-1/h+EY_n}^{0} |x|^3 dF(x) + |EY_n|^3 P\left(|X| \geq \frac{1}{h} + EY_n\right)\right)$$

$$= o\left(h^{2-k} G\left(\frac{1}{h}\right)\right)$$

for $k = 0, 1, 2$. It follows that

$$\hat{L}_{k,n} = \frac{h^{2-k}}{(2-k)!} G\left(\frac{1}{h}\right)(1 + o(1)), \quad k = 0, 1, 2.$$

The latter and the above bounds for $\bar{L}_{k,n}$ yield that

$$L_{k,n} = \frac{h^{2-k}}{(2-k)!} G\left(\frac{1}{h}\right)(1 + o(1)), \quad k = 0, 1, 2. \tag{2.41}$$

Relations (2.32), (2.35) and (2.41) imply (2.28). Assertions (2.28), (2.33), (2.35) and (2.41) yield (2.29). From (2.28), (2.29), (2.34), (2.35) and (2.41), we obtain (2.30). Relation (2.31) follows from (2.28) and (2.29).

If condition 1) holds, the proof follows the same pattern. One has to put $y_n = \infty$ for all n. The above calculations change evidently in this case. We omit details. □

The next result shows that one can replace (2.27) by simpler conditions.

Lemma 2.15. *Assume that* $EX = 0$, $E(X^+)^2 < \infty$ *and* $F \in D(2)$. *Suppose that* $y = y_n \nearrow \infty$, $h = h_n \to 0$ *and one of the following conditions holds:*

1) there exists $\beta \in (0, 1)$ *such that* $\limsup h_n y_n^{1-\beta} < 1$ *and* $Ee^{(X^+)^\beta} < \infty$.

2) $h_n y_n = O(1)$.

Then relations (2.28)–(2.31) hold.

The proof of Lemma 2.15 is similar to that of Lemma 2.10. We need only prove that (2.27) follows from conditions 1) or 2). To this end, we replace x^3 by x^2 in the definition of the integral J_n and note that bounds for the integral I_n require the condition $E(X^+)^2 < \infty$. Changes in the proof of Lemma 2.10 are obvious and we omit details.

2.10 Asymptotic Expansions of Functions of Large Deviations Theory for $DN(\alpha)$ and $D(\alpha)$

In this section, we find asymptotic expansions for the LDT functions when $F \in D(\alpha)$ and $F \in DN(\alpha)$, $\alpha \in (1,2)$. The cases of finite and infinite exponential moments are included both in the next result.

Lemma 2.16. *Put $G(x) = x^\alpha F(-x)$ for $x > 0$. (Here $G(x) \in SV_\infty$.) Assume that $EX = 0$, $h = h_n \to 0$ and one of the following conditions holds:*

1) $h_0 > 0$ and $y = y_n = \infty$ for all n,

2) $y = y_n \to \infty$ and

$$h^{2-\alpha} \int_0^\infty x^2 e^{hx} dF_n(x) = o\left(G\left(\frac{1}{h}\right)\right). \tag{2.42}$$

If $F \in D(\alpha)$, $\alpha \in (1,2)$, then

$$\varphi_n(h) = 1 + c_1 h^\alpha G\left(\frac{1}{h}\right)(1+o(1)), \tag{2.43}$$

$$m_n(h) = c_2 h^{\alpha-1} G\left(\frac{1}{h}\right)(1+o(1)), \tag{2.44}$$

$$\sigma_n^2(h) = c_3 h^{\alpha-2} G\left(\frac{1}{h}\right)(1+o(1)), \tag{2.45}$$

$$f_n(h) = c_4 h^\alpha G\left(\frac{1}{h}\right)(1+o(1)), \tag{2.46}$$

where $c_1 = \Gamma(2-\alpha)/(\alpha-1)$, $c_2 = \alpha\Gamma(2-\alpha)/(\alpha-1)$, $c_3 = \alpha\Gamma(2-\alpha)$ and $c_4 = \Gamma(2-\alpha)$. Here $\Gamma(x)$ is the gamma-function.

If $F \in DN(\alpha)$, $\alpha \in (1,2)$, then

$$\varphi_n(h) = 1 + \frac{h^\alpha}{\alpha}(1+o(1)), \tag{2.47}$$

$$m_n(h) = h^{\alpha-1}(1+o(1)), \tag{2.48}$$

$$\sigma_n^2(h) = (\alpha-1)h^{\alpha-2}(1+o(1)), \tag{2.49}$$

$$f_n(h) = \frac{\alpha-1}{\alpha} h^\alpha (1+o(1)). \tag{2.50}$$

Note that if condition 1) holds then $\varphi_n(h) = \varphi(h)$, $m_n(h) = m(h)$, $\sigma_n^2(h) = \sigma^2(h)$ and $f_n(h) = f(h)$ for $h \in (0, h_0)$ and all n.

Proof. Assume that condition 2) holds.

Suppose first that $F \in D(\alpha)$, $\alpha \in (1, 2)$. By Theorem 2.1, we get $G(x) \in SV_\infty$.

Let $L_{k,n}$ and $I_{k,n}$ be the integrals from the proof of Lemma 2.13. We write (2.32)–(2.34) and investigate the behaviour of $L_{k,n}$ and $I_{k,n}$. By Lemma 2.9 and (2.42), we have

$$I_{k,n} \le h^{2-k} \int_0^\infty x^2 e^{hx} dF_n(x) = o\left(h^{\alpha-k} G\left(\frac{1}{h}\right)\right), \quad k = 0, 1, 2. \quad (2.51)$$

Put $r_k(x) = x^k e_{1-k}(x)$, $k = 0, 1, 2$. Then an integration by parts yields

$$L_{k,n} = h^{-k} \int_{-\infty}^0 r_k(hx) dF_n(x) = h^{-k+1} \int_{-\infty}^0 F_n(x) r_k'(hx) dx$$

$$= h^{-k} \int_0^\infty F_n\left(-\frac{x}{h}\right) r_k'(-x) dx.$$

We now check that

$$\int_1^\infty F_n\left(-\frac{x}{h}\right) r_k'(-x) dx \sim h^\alpha G\left(\frac{1}{h}\right) \int_1^\infty x^{-\alpha} r_k'(-x) dx, \quad k = 0, 1, 2. \quad (2.52)$$

Since $F_n(-x) = F(-x + EY_n)$ for $x \ge 0$, $EY_n < 0$ and $EY_n \to 0$, we have

$$\frac{h^\alpha}{(\rho x)^\alpha} G\left(\frac{\rho x}{h}\right) = F\left(-\frac{\rho x}{h}\right) \le F_n\left(-\frac{x}{h}\right) \le F\left(-\frac{x}{h}\right) = \frac{h^\alpha}{x^\alpha} G\left(\frac{x}{h}\right)$$

for every $\rho > 1$, all $x \ge \varepsilon$ and all sufficiently large n.

In view of $r_0'(-x) = e^{-x} - 1 \le 0$ and $r_1'(-x) = e^{-x} - 1 - xe^{-x} \le 0$ for $x \ge 0$, it follows by Theorem 2.6 on p. 64 in [Seneta (1976)] that

$$\int_1^\infty F_n\left(-\frac{x}{h}\right) r_k'(-x) dx \ge h^\alpha \int_1^\infty G\left(\frac{x}{h}\right) x^{-\alpha} r_k'(-x) dx$$

$$\sim h^\alpha G\left(\frac{1}{h}\right) \int_1^\infty x^{-\alpha} r_k'(-x) dx$$

for $k = 0, 1$. By the same argument, we have

$$\int_1^\infty F_n\left(-\frac{x}{h}\right) r_k'(-x)dx \le \frac{h^\alpha}{\rho^\alpha} \int_1^\infty G\left(\frac{\rho x}{h}\right) x^{-\alpha} r_k'(-x)dx$$

$$\sim \frac{h^\alpha}{\rho^\alpha} G\left(\frac{1}{h}\right) \int_1^\infty x^{-\alpha} r_k'(-x)dx$$

for $k = 0, 1$. Since ρ may be chosen arbitrary close to 1, we arrive at (2.52) for $k = 0, 1$.

Further, $r_2'(-x) = x(x-2)\mathrm{e}^{-x}$. Hence $r_2'(-x) \le 0$ for $x \in (1, 2)$ and $r_2'(-x) \ge 0$ for $x \ge 2$. In the same way as before, one can check that

$$\int_2^\infty F_n\left(-\frac{x}{h}\right) r_2'(-x)dx \sim h^\alpha G\left(\frac{1}{h}\right) \int_2^\infty x^{-\alpha} r_2'(-x)dx$$

and

$$\int_1^2 F_n\left(-\frac{x}{h}\right) r_2'(-x)dx \sim h^\alpha G\left(\frac{1}{h}\right) \int_1^2 x^{-\alpha} r_2'(-x)dx.$$

The latter two relations imply (2.52) for $k = 2$.

We now prove that

$$\int_0^1 F_n\left(-\frac{x}{h}\right) r_k'(-x)dx \sim h^\alpha G\left(\frac{1}{h}\right) \int_0^1 x^{-\alpha} r_k'(-x)dx, \quad k = 0, 1, 2. \tag{2.53}$$

It is clear that for every $\rho > 1$ and all sufficiently large n the following inequalities

$$\int_{\rho-1}^1 F\left(-\frac{\rho x}{h}\right) r_k'(-x)dx \ge \int_0^1 F_n\left(-\frac{x}{h}\right) r_k'(-x)dx \ge \int_0^1 F\left(-\frac{x}{h}\right) r_k'(-x)dx$$

hold for $k = 0, 1, 2$. Theorem 2.7 on p. 66 in [Seneta (1976)] yields

$$\int_0^1 F\left(-\frac{x}{h}\right) r_k'(-x)dx \sim h^\alpha G\left(\frac{1}{h}\right) \int_0^1 x^{-\alpha} r_k'(-x)dx, \quad k = 0, 1, 2.$$

By relation (2.11) on p. 63 in [Seneta (1976)], we have

$$\int_{\rho-1}^1 F\left(-\frac{x}{h}\right) r_k'(-x)dx \sim h^\alpha G\left(\frac{1}{h}\right) \int_{\rho-1}^1 x^{-\alpha} r_k'(-x)dx, \quad k = 0, 1, 2.$$

Since ρ may be chosen arbitrarily close to 1, the last two relations yield (2.53).

It follows from (2.52) and (2.53) that

$$L_{k,n} = c_{k+1} h^{\alpha-k} G\left(\frac{1}{h}\right)(1+o(1)), \quad k=0,1,2, \tag{2.54}$$

where

$$c_1 = \int_0^\infty x^{-\alpha}(1-\mathrm{e}^{-x})dx = \frac{\Gamma(2-\alpha)}{\alpha-1},$$

$$c_2 = \int_0^\infty x^{-\alpha}(1-\mathrm{e}^{-x}+x\mathrm{e}^{-x})dx = \frac{\alpha\Gamma(2-\alpha)}{\alpha-1},$$

$$c_3 = \int_0^\infty x^{-\alpha}(2x-x^2)\mathrm{e}^{-x}dx = \alpha\Gamma(2-\alpha).$$

Relations (2.32), (2.51) and (2.54) imply (2.43). Assertions (2.43), (2.33), (2.51) and (2.54) yield (2.44). From (2.44), (2.43), (2.34), (2.51) and (2.54), we get (2.45). Relation (2.46) follows from (2.43) and (2.44).

Finally, we assume that $F \in DN(\alpha)$, $\alpha \in (1,2)$. The proof is the same as that above. We need only mention that Theorem 2.6.7 and the last two formulae before the case (3) on p. 22 in [Ibragimov and Linnik (1971)] yield that

$$F(-x) = x^{-\alpha}\frac{\alpha-1}{\alpha\Gamma(2-\alpha)}(1+o(1)) \quad \text{as} \quad x \to \infty.$$

It follows that

$$G\left(\frac{1}{h}\right) = \frac{\alpha-1}{\alpha\Gamma(2-\alpha)}(1+o(1)).$$

Hence, we get (2.47)–(2.50) instead of (2.43)–(2.46).

If condition 1) holds, the proof follows the same pattern. One has to put $y_n = \infty$ for all n. The above calculations change evidently in this case. We omit details. □

The next result allows to replace condition (2.42) by simpler ones.

Lemma 2.17. *Assume that $EX = 0$, $y = y_n \to \infty$, $h = h_n \to 0$ and one of the following conditions holds:*

1) there exists $\beta \in (0,1)$ such that $\limsup h_n y_n^{1-\beta} < 1$ and $E\mathrm{e}^{(X^+)^\beta} < \infty$.

2) $h_n y_n = O(1)$.

If $F \in D(\alpha)$, $\alpha \in (1,2)$ *then relations (2.43)–(2.46) hold. If* $F \in DN(\alpha)$, $\alpha \in (1,2)$ *then relations (2.47)–(2.50) hold.*

Proof. We need the next result that is a consequence of the asymmetry of stable laws under consideration.

Lemma 2.18. *If* $F \in DN(\alpha)$ *or* $F \in D(\alpha)$, $\alpha \in (1,2)$, *then* $1 - F(x) = o(F(-x))$ *and*

$$\int_0^x u^2 dF(u) = o\left(x^{2-\alpha} G(x)\right) \quad as \quad x \to \infty.$$

Proof. From [Feller (1971)], we have the first relation and $\mu(x) \sim x^{2-\alpha} G(x)$ as $x \to \infty$, where $\mu(x) = \mu^+(x) + \mu^-(x)$,

$$\mu^-(x) = \int_{-x}^0 u^2 dF(u), \quad \mu^+(x) = \int_0^x u^2 dF(u).$$

An integrating by parts gives

$$2\int_0^x u(1 - F(u))du = \mu^+(x) + x^2(1 - F(x)),$$

$$2\int_0^x uF(-u)du = \mu^-(x) + x^2 F(-x).$$

Take $\varepsilon > 0$. Then $1 - F(x) \leq \varepsilon F(-x)$ for all $x \geq x_0 = x_0(\varepsilon)$ and

$$\int_0^x u(1 - F(u))du \leq x_0^2 + \varepsilon \int_{x_0}^x uF(-u)du \leq 2\varepsilon \int_0^x uF(-u)du$$

for all sufficiently large x. Here we have used that the last integral tends to infinity. Hence,

$$\mu^+(x) = o(\mu^-(x) + x^2 F(-x)) = o(\mu(x) + x^2 F(-x))$$

as $x \to \infty$ and the second relation follows. □

The proof of Lemma 2.17 is similar to that of Lemma 2.10.

Assume that condition 2) holds. Then

$$J_n = \int_0^\infty x^2 \mathrm{e}^{hx} dF_n(x) = \int_0^{z_n} x^2 \mathrm{e}^{hx} dF_n(x + EY_n) + z_n^2 \mathrm{e}^{hz_n} P(X \geq y).$$

In the same way as in Lemma 2.10, we obtain

$$I_n = \int_0^{z_n} x^2 e^{hx} dF_n(x + EY_n) \le C \int_0^{y} x^2 e^{hx} dF(x).$$

Taking into account that $y \le C_1/h$ and Lemma 2.18, we have $I_n = o(h^{\alpha-2}G(1/h))$. Applying again Lemma 2.18, we get

$$z_n^2 e^{hz_n} P(X \ge y) = o\left(y^2 F(-y)\right) = o\left(y^{2-\alpha} G(y)\right).$$

By [Seneta (1976)], there exists a non-decreasing function $R(x) \in RV_\infty$ such that $R(y) \sim y^{2-\alpha}G(y)$. Since $R(y) \le R(C_1/h) \sim C_1^{2-\alpha}R(1/h)$, we have that

$$z_n^2 e^{hz_n} P(X \ge y) = o\left(h^{\alpha-2} G\left(\frac{1}{h}\right)\right).$$

Hence, condition (2.42) holds and the result follows from Lemma 2.16.

If condition 1) is satisfied, then calculations are similar to those in Lemma 2.10. One has to replace x^3 by x^2 only. Then we get $J_n = O(1)$ which yields (2.42) in view of $G(x) \in SV_\infty$. □

2.11 Large Deviations for $D(2)$

In this section, we discuss the asymptotic behaviour of the large deviation probabilities for random variables from the domain of non-normal attraction of the normal law.

We will use the following notations. For $g(x) \in RV_\infty$ ($g(x) \in RV_0$), let $g^{-1}(x)$ be an asymptotically inverse function, i.e. $g(g^{-1}(x)) \sim g^{-1}(g(x)) \sim x$ as $x \to \infty$ ($x \to 0$). The asymptotically inverse function is asymptotically unique. One can find details in [Seneta (1976)] for RV_∞. We consider the case RV_0 below.

Our first result is for random variables satisfying the one-sided Cramér condition.

Theorem 2.6. *Assume that* $h_0 > 0$, $EX = 0$ *and* $F \in D(2)$. *For* $h > 0$ *and* $x > 0$, *put*

$$\hat{m}(h) = hG\left(\frac{1}{h}\right), \quad \hat{f}(h) = \frac{h^2}{2}G\left(\frac{1}{h}\right), \quad \hat{\gamma}(x) = \hat{m}\left(\hat{f}^{-1}(x)\right) \tag{2.55}$$

where $G(x) = \int_{-x}^{0} u^2 dF(u)$. *Then for every sequence* $\{x_n\}$ *with* $x_n \to 0$ *and* $nx_n \to \infty$, *the following relation holds*

$$\log P(S_n \ge n\hat{\gamma}(x_n)) = -nx_n(1 + o(1)). \tag{2.56}$$

The functions $\hat{m}(h)$, $\hat{f}(h)$ and $\hat{\gamma}(x)$ are the main terms of asymptotic expansions for $m(h)$, $f(h)$ and $\gamma(x)$ at zero, correspondingly. Moreover, $G(x) \in SV_\infty$, $\hat{m}(h) \in RV_0$ and $\hat{f}(h) \in RV_0$.

In view of an obvious replacement, the conclusion of Theorem 2.6 means that

$$\log P\left(S_n \geq n\hat{\gamma}\left(\frac{x_n^2}{2n}\right)\right) \sim -\frac{x_n^2}{2} \tag{2.57}$$

for every sequence $\{x_n\}$ with $x_n \to \infty$ and $x_n = o(\sqrt{n})$. Note that $F \in DN(2)$ formally corresponds to the case $G(x) = 1$ for all x for which $\hat{\gamma}(x) = \sqrt{2x}$ and (2.57) turns to (2.21) of Theorem 2.3.

Proof. We first prove three lemmas.

Lemma 2.19. *For every sequence $\{x_n\}$ of positive numbers, one has*

$$\gamma_n(x_n(1+o(1))) = \gamma_n(x_n)(1+o(1)),$$
$$\zeta_n(x_n(1+o(1))) = \zeta_n(x_n)(1+o(1)).$$

Proof. Let $\{\varepsilon_n\}$ be a sequence of real numbers such that $\varepsilon_n \to 0$. The concavity and monotonicity of $\gamma_n(x)$ and $\gamma_n(0) = 0$ yield that

$$(1 - |\varepsilon_n|)\gamma_n(x_n) \leq \gamma_n(x_n(1+\varepsilon_n)) \leq (1 + |\varepsilon_n|)\gamma_n(x_n)$$

for all n. This implies the first relation. The second one follows by the same way in view of the convexity and monotonicity of $\zeta_n(x)$ and $\zeta_n(0) = 0$. □

Lemma 2.20. *Assume that $f(x) = x^\alpha G(1/x)$, where $G(x) \in SV_\infty$. Then there exists a function $L(x) \in SV_\infty$ such that the function $f^{-1}(x) = x^{1/\alpha}L(1/x)$ is the asymptotically inverse function to $f(x)$. Moreover, if $f_{-1}(x)$ is another asymptotically inverse function then $f_{-1}(x) \sim f^{-1}(x)$ as $x \to 0$.*

Proof. By the representation of a slowly varying function (see Theorem 1.2 on p.2 [Seneta (1976)]), there exist functions $\eta(t)$, $\eta(t) \to const$ as $t \to \infty$, and $\epsilon(t)$, $\epsilon(t) \to 0$ as $t \to \infty$, and a positive constant B such that

$$f(x) = \exp\Big\{\eta(1/x) + \int_B^{1/x} \frac{\epsilon(t) - \alpha}{t} dt\Big\} \sim C\exp\Big\{-\int_B^{1/x} \frac{\alpha - \epsilon(t)}{t} dt\Big\} = Cr_1(x),$$

as $x \to 0$. We further assume that $x \leq 1/B$ and $\alpha - \epsilon(t) > 0$ for $t \geq B$.

The function $r_1(x)$ is continuous, strictly increasing on $[0, 1/B]$ and $r_1(0) = 0$. Let $r_2(x)$ be the inverse function to $r_1(x)$. Hence, $r_2(x)$ is continuous, strictly increasing on $[0, r_1(1/B)]$ and $r_2(0) = 0$. We have from $r_2(r_1(x)) = r_1(r_2(x)) = x$ that $r_2'(r_1(x))r_1'(x) = r_1'(r_2(x))r_2'(x) = 1$. This yields that

$$\frac{r_2'(r_1(x))r_1(x)}{x} = \frac{1}{\alpha - \epsilon(1/x)}.$$

Putting $x = r_2(t)$ in the last relation, we have

$$\frac{r_2'(t)t}{r_2(t)} = \frac{1}{\alpha - \epsilon(1/r_2(t))} = \frac{1}{\alpha} + \frac{\alpha^{-2}\epsilon(1/r_2(t))}{1 - \alpha^{-1}\epsilon(1/r_2(t))} = \frac{1}{\alpha} + \epsilon_1(t)$$

for all $t \le B_1 = r_1(1/B)$, where $\epsilon_1(t) \to 0$ as $t \to 0$. It follows that

$$r_2(x) = \exp\left\{-\int_x^{B_1} \frac{\alpha^{-1} + \epsilon_1(t)}{t} dt\right\} = \exp\left\{-\int_{1/B_1}^{1/x} \frac{\alpha^{-1} + \epsilon_1(1/u)}{u} du\right\}$$

$$= x^{1/\alpha} \exp\left\{B_2 - \int_{1/B_1}^{1/x} \frac{\epsilon_2(u)}{u} du\right\} = x^{1/\alpha}\ell(1/x).$$

Here $\ell(t) \in SV_\infty$ by the representation of a slowly varying function.

Put $f^{-1}(x) = x^{1/\alpha}L(1/x)$, where $L(x) = C^{-1/\alpha}\ell(x)$. By the uniform convergence theorem (Theorem 1.1 on p.2 in [Seneta (1976)]), we have

$$f(f^{-1}(x)) \sim Cr_1(C^{-1/\alpha}r_2(x)) = Cr_1(r_2(C^{-1}x(1 + o(1)))) \sim x,$$

and

$$\begin{aligned} f^{-1}(f(x)) &= C^{-1/\alpha}r_2(Cr_1(x)(1 + o(1))) \\ &= C^{-1/\alpha}r_2(r_1(C^{1/\alpha}x(1 + o(1)))) \sim x, \end{aligned}$$

as $x \to 0$.

Check that $f^{-1}(x)$ is asymptotically unique. For another asymptotically inverse function $f_{-1}(x)$, the uniform convergence theorem yields

$$f_{-1}(x) \sim f^{-1}(f(f_{-1}(x))) = f^{-1}(x(1 + o(1))) \sim f^{-1}(x)$$

as $x \to 0$. □

Lemma 2.21. *If $\hat{m}(h)$, $\hat{f}(h)$ and $\hat{\gamma}(x)$ are defined by formulae (2.55), then there exists a function $L(x) \in SV_\infty$ such that*

$$\hat{f}^{-1}(x) = \sqrt{2x}L\left(\frac{1}{x}\right), \tag{2.58}$$

$$\hat{\gamma}(x) \sim \frac{\sqrt{2x}}{L(1/x)} \quad as \quad x \to 0. \tag{2.59}$$

Proof. An application of Lemma 2.20 yields relation (2.58). This and (2.55) imply that

$$\hat{\gamma}(x) = \sqrt{2x}L\left(\frac{1}{x}\right) G\left(\frac{1}{\sqrt{2x}L\,(1/x)}\right).$$

In view of

$$x \sim \hat{f}(\hat{f}^{-1}(x)) = x\left(L\left(\frac{1}{x}\right)\right)^2 G\left(\frac{1}{\sqrt{2x}L\,(1/x)}\right) \quad \text{as} \quad x \to 0,$$

we obtain (2.59). □

Note that $L(x) \to 0$ as $x \to \infty$ in Theorem 2.6 because $G(x) \to \infty$ as $x \to \infty$.

Put $y_n = \infty$ and $h_n = \hat{f}^{-1}(x_n)$ for all n. By Lemmas 2.13 and 2.19 and the definition of $\gamma(x)$, we have

$$\begin{aligned}\gamma(x_n) &= \gamma(\hat{f}(h_n)) = \gamma(f(h_n)(1 + o(1))) \\ &\sim \gamma(f(h_n)) = m(h_n) \sim \hat{m}(h_n) \sim \hat{\gamma}(x_n).\end{aligned}$$

By Lemma 2.2, we get

$$\log P(S_n \geq n\gamma(x_n)) \leq -nx_n.$$

Take $\varepsilon > 0$. We have by (2.59) that $\hat{\gamma}((1+\varepsilon)x_n) \sim \sqrt{1+\varepsilon}\hat{\gamma}(x_n)$. Then $\gamma(x_n) \leq \hat{\gamma}((1+\varepsilon)x_n)$ for all sufficiently large n. It yields that

$$P(S_n \geq n\hat{\gamma}((1+\varepsilon)x_n)) \leq P(S_n \geq n\gamma(x_n))$$

for all sufficiently large n. Hence,

$$\log P(S_n \geq n\hat{\gamma}((1+\varepsilon)x_n)) \leq -nx_n$$

for all sufficiently large n. Replacing $(1+\varepsilon)x_n$ by x_n, we have

$$\log P(S_n \geq n\hat{\gamma}(x_n)) \leq -n\frac{x_n}{1+\varepsilon}$$

for all sufficiently large n. The latter implies that

$$\limsup \frac{1}{nx_n} \log P(S_n \geq n\hat{\gamma}(x_n)) \leq -\frac{1}{1+\varepsilon}.$$

Passing to the limit as $\varepsilon \to 0$ in the last inequality, we obtain

$$\limsup \frac{1}{nx_n} \log P(S_n \geq n\hat{\gamma}(x_n)) \leq -1. \tag{2.60}$$

By Lemma 2.13, the relations $\sqrt{n}\sigma(h_n) = o(nm(h_n))$ and $h_n\sqrt{n}\sigma(h_n) = o(nf(h_n))$ hold.

Take $\varepsilon \in (0,1)$. Since $\hat{\gamma}((1-\varepsilon)x_n) \sim \sqrt{1-\varepsilon}\hat{\gamma}(x_n)$, the inequality $nm(h_n) - 2\sqrt{n}\sigma(h_n) \geq n\hat{\gamma}((1-\varepsilon)x_n)$ holds for all sufficiently large n. Hence,

$$P(S_n \geq nm(h_n) - 2\sqrt{n}\sigma(h_n)) \leq P(S_n \geq n\hat{\gamma}((1-\varepsilon)x_n))$$

for all sufficiently large n. Lemma 2.2 implies that

$$\log P(S_n \geq n\hat{\gamma}((1-\varepsilon)x_n)) \geq -nx_n(1+\varepsilon)$$

for all sufficiently large n. Replacing $(1-\varepsilon)x_n$ by x_n, we get

$$\log P(S_n \geq n\hat{\gamma}(x_n)) \geq -nx_n\frac{1+\varepsilon}{1-\varepsilon}$$

for all sufficiently large n. It follows that

$$\liminf \frac{1}{nx_n} \log P(S_n \geq n\hat{\gamma}(x_n)) \geq -\frac{1+\varepsilon}{1-\varepsilon}.$$

Passing to the limit as $\varepsilon \to 0$, we arrive at

$$\liminf \frac{1}{nx_n} \log P(S_n \geq n\hat{\gamma}(x_n)) \geq -1.$$

The last relation and (2.60) yield (2.56). □

We deal with the one-sided Linnik condition in the next result.

Theorem 2.7. *Assume that* $Ee^{(X^+)^\beta} < \infty$, $EX = 0$ *and* $F \in D(2)$. *Define* $\hat{m}(h)$, $\hat{f}(h)$ *and* $\hat{\gamma}(x)$ *by formulae (2.55). Let* $L(x)$ *be the function from (2.58). Then relation (2.56) holds for every sequence* $\{x_n\}$ *with* $nx_n \to \infty$ *and* $n^{1-\beta}x_n^{(2-\beta)/2}(L(1/x_n))^\beta \to 0$.

Writing (2.56) in form of (2.57), we see that Theorem 2.7 corresponds to Theorem 2.4 in which $F \in DN(2)$ and $G(x) = L(x) = 1$ for all x.

Proof. The second condition on x_n yields that $x_n \to 0$ in Theorem 2.7.

Put $y = y_n = n\hat{m}(\hat{f}^{-1}(x_n))$ and $h = h_n = \hat{f}^{-1}(x_n)$. Relations (2.58) and (2.59) imply that $y \to \infty$, $h \to 0$ and $hy^{1-\beta} = o(1)$. By Lemma 2.15, relations (2.28)–(2.31) hold.

By Lemmas 2.15 and 2.19 and the definition of $\gamma_n(x)$, we have

$$\begin{aligned}\gamma_n(x_n) &= \gamma_n(\hat{f}(h_n)) = \gamma_n(f_n(h_n)(1+o(1)))\\ &\sim \gamma_n(f_n(h_n)) = m_n(h_n) \sim \hat{m}(h_n) \sim \hat{\gamma}(x_n).\end{aligned}$$

We will apply Lemma 2.4 with $k = n$ and $x = \hat{\gamma}(x_n)$. We have by (2.31) that

$$\zeta_n(x_n) = \hat{\zeta}(x) \sim \hat{f}(h) = x_n.$$

By relation (2.59), we get

$$(n\hat{\gamma}(x_n))^\beta n^{-\beta/2} \sim \frac{(2nx_n)^{\beta/2}}{(L(1/x_n))^\beta} \to \infty$$

and

$$\frac{(n\hat{\gamma}(x_n))^\beta}{nx_n} \sim 2^{\beta/2} \frac{n^{\beta-1}x_n^{(\beta-2)/2}}{(L(1/x_n))^\beta} \to \infty.$$

These relations and the one-sided Linnik condition imply that

$$kP(X \geq y) = o\left(n\mathrm{e}^{-(n\hat{\gamma}(x_n))^\beta}\right) = o\left(\mathrm{e}^{-(n\hat{\gamma}(x_n))^\beta(1+o(1))}\right) = o\left(\mathrm{e}^{-2nx_n}\right) = o(1).$$

Therefore

$$kP(X \geq y)\mathrm{e}^{-k\hat{\zeta}(\delta x)} + (kP(X \geq y))^2 \leq 2kP(X \geq y) = o(\mathrm{e}^{-2nx_n}),$$

and condition (2.8) holds with $H_1 = 1$ for all sufficiently large n. It follows by Lemma 2.4 that

$$P(S_k \geq kx) \leq 2\mathrm{e}^{-k\hat{\zeta}(x)} + kP(X \geq (1-\delta)kx)$$

for all sufficiently large n. Taking again into account the asymptotic behaviour of $(n\hat{\gamma}(x_n))^\beta$ and the Linnik condition, we have

$$kP(X \geq (1-\delta)kx) = o\left(n\mathrm{e}^{-((1-\delta)n\hat{\gamma}(x_n))^\beta}\right) = o\left(\mathrm{e}^{-2nx_n}\right).$$

This implies that

$$P(S_k \geq kx) \leq 3\mathrm{e}^{-k\hat{\zeta}(x)}$$

for all sufficiently large n. Hence, for every $\varepsilon \in (0,1)$, the inequality

$$P(S_n \geq n\hat{\gamma}(x_n)) \leq 3\mathrm{e}^{-(1-\varepsilon)nx_n}$$

holds for all sufficiently large n. It follows that

$$\limsup \frac{1}{nx_n} \log P(S_n \geq n\hat{\gamma}(x_n)) \leq -(1-\varepsilon).$$

Passing to the limit as $\varepsilon \to 0$, we conclude that

$$\limsup \frac{1}{nx_n} \log P(S_n \geq n\hat{\gamma}(x_n)) \leq -1. \tag{2.61}$$

Now we check that the conditions of Lemma 2.5 hold. Take $\delta \in (0,1)$. We have

$$E\hat{X} = -\int_y^\infty u dF(u) + yP(X \geq y) = o\left(y\mathrm{e}^{-y^\beta}\right) = o\left(\mathrm{e}^{-(1-\delta)(n\hat{\gamma}(x_n))^\beta}\right).$$

This and assertions (2.30) and (2.31) imply that $-E\hat{X} = o(x)$, $h\sigma_n(h) = o(\sqrt{n} f_n(h))$, and the conditions of Lemma 2.5 are satisfied.

Take $\tau \in (0,1)$. Since $n\hat{\gamma}((1-\tau)^3 x_n) \sim (1-\tau)^{3/2} n\hat{\gamma}(x_n) \geq (1-\tau)^2 n\hat{\gamma}(x_n)$ for all sufficiently large n, by Lemma 2.5, the inequalities

$$P(S_n \geq n\hat{\gamma}((1-\tau)^3 x_n)) \geq P(S_k \geq (1-\tau)^2 kx) \geq e^{-k\hat{\zeta}(x)(1+\tau)}$$

hold for all sufficiently large n. Then

$$\liminf \frac{1}{nx_n} \log P(S_n \geq n\hat{\gamma}((1-\tau)^3 x_n)) \geq -(1+\tau).$$

Replacing $(1-\tau)^3 x_n$ by x_n in the last inequality, we have

$$\liminf \frac{1}{nx_n} \log P(S_n \geq n\hat{\gamma}(x_n)) \geq -\frac{1+\tau}{(1-\tau)^3}.$$

Passing to the limit as $\tau \to 0$ in the latter inequality and taking into account (2.61), we get (2.56). □

In the following result, we consider a power moment condition.

Theorem 2.8. *Assume that* $EX = 0$, $E(X^+)^p < \infty$ *for some* $p > 2$ *and* $F \in D(2)$. *Define* $\hat{m}(h)$, $\hat{f}(h)$ *and* $\hat{\gamma}(x)$ *by formulae (2.55). Let* $L(x)$ *be the function from (2.58). Then relation (2.56) holds for every sequence* $\{x_n\}$ *with* $nx_n \to \infty$ *and* $\limsup(2nx_n)/\log n \leq p-2$.

Writing relation (2.56) in form of (2.57), we conclude that Theorem 2.8 corresponds to Theorem 2.5 where $F \in DN(2)$ and $G(x) = L(x) = 1$ for all x.

Proof. Note that $x_n \to 0$ in the assumptions of Theorem 2.8.

Put $h = h_n = \hat{f}^{-1}(x_n)$ and $y = y_n = 1/h_n$ for all n. By (2.58), we have $h \to 0$ and $y \to \infty$. Lemma 2.15 implies that relations (2.28)–(2.31) hold.

Put $k = n$ and $x = \hat{\gamma}(x_n)$. We will check that the conditions of Lemma 2.4 hold.

For $\delta \in (0,1]$, we have

$$\hat{\zeta}(\delta x) = \zeta_n(\delta\hat{\gamma}(x_n)) = \zeta_n(\hat{\gamma}(\delta^2 x_n)(1+o(1))) = \zeta_n(\gamma_n(\delta^2 x_n(1+o(1))) \sim \delta^2 x_n.$$

It follows by Lemma 2.12 that for every $\varepsilon \in (0,1)$,

$$n^{1-p} x_n^{-p/2} = O\left(e^{-nx_n(1-\varepsilon)}\right).$$

The condition $E(X^+)^p < \infty$ yields that

$$kP(X \geq y) = o\left(nx_n^{p/2}\left(L\left(\frac{1}{x_n}\right)\right)^p\right) = o\left((nx_n)^p e^{-nx_n(1-\varepsilon)}\right) = o(1).$$

It follows that

$$kP(X \geq y)\mathrm{e}^{-k\hat{\zeta}(\delta x)} + (kP(X \geq y))^2 =$$
$$= o\left((nx_n)^p \mathrm{e}^{-(1-\varepsilon+\delta^2)nx_n}\right) + o\left((nx_n)^{2p}\mathrm{e}^{-2nx_n(1-\varepsilon)}\right) = o\left(\mathrm{e}^{-nx_n}\right),$$

provided $\delta^2 > \varepsilon$.

Hence, the conditions of Lemma 2.4 hold for all sufficiently large n. Then

$$P(S_k \geq kx) \leq 2\mathrm{e}^{-nx_n} + kP(X \geq (1-\delta)kx)$$

for all sufficiently large n. Making use of $E(X^+)^p < \infty$ again, we have

$$kP(X \geq (1-\delta)kx) = o\left(n^{1-p}x_n^{-p/2}\left(L\left(\frac{1}{x_n}\right)\right)^p\right) = o\left(\mathrm{e}^{-nx_n(1-\varepsilon)}\right).$$

This implies that

$$P(S_k \geq kx) \leq 3\mathrm{e}^{-nx_n(1-\varepsilon)}$$

for all sufficiently large n.

The rest of the proof for the upper bound coincides with that of the previous theorem. We omit details and turn to the lower bound. To this end, we check the conditions of Lemma 2.5.

We have

$$E\hat{X} = -\int_y^\infty u dF(u) + yP(X \geq y) = o(y^{1-p}) = o\left(x_n^{(p-1)/2}\left(L\left(\frac{1}{x_n}\right)\right)^{p-1}\right).$$

The latter implies that $-E\hat{X} = o(x)$. By $h\sigma_n(h) = o(\sqrt{n}f_n(h))$, we conclude that the conditions of Lemma 2.5 hold. The remainder of the proof for the lower bound is the same as that of Theorem 2.7. Details are omitted. □

2.12 Large Deviations for $DN(\alpha)$ and $D(\alpha)$

We turn now to large deviations for $F \in DN(\alpha)$ and $F \in D(\alpha)$ with $\alpha \in (1,2)$. We again start with random variables satisfying the Cramér condition.

Theorem 2.9. *Assume that $h_0 > 0$, $EX = 0$ and $\alpha \in (1,2)$.*

If $F \in D(\alpha)$, then for $h > 0$ and $x > 0$, put

$$\hat{m}(h) = c_2 h^{\alpha-1} G\left(\frac{1}{h}\right), \quad \hat{f}(h) = c_4 h^\alpha G\left(\frac{1}{h}\right), \quad \hat{\gamma}(x) = \hat{m}\left(\hat{f}^{-1}(x)\right), \quad (2.62)$$

where $G(x) = x^{\alpha}F(-x)$, $c_2 = \Gamma(2-\alpha)/(\alpha-1)$ *and* $c_4 = \Gamma(2-\alpha)$.

If $F \in DN(\alpha)$, *then put*

$$\hat{m}(h) = h^{\alpha-1}, \quad \hat{f}(h) = \frac{\alpha-1}{\alpha}h^{\alpha}, \quad \hat{\gamma}(x) = \left(\frac{x}{\lambda}\right)^{\lambda} \tag{2.63}$$

for $h > 0$ *and* $x > 0$, *where* $\lambda = (\alpha-1)/\alpha$.

Then for every sequence $\{x_n\}$ *with* $x_n \to 0$ *and* $nx_n \to \infty$, *the following relation holds*

$$\log P(S_n \geq n\hat{\gamma}(x_n)) = -nx_n(1+o(1)). \tag{2.64}$$

Note that the functions $\hat{m}(h)$, $\hat{f}(h)$ and $\hat{\gamma}(x)$ are the main terms of asymptotic expansions for $m(h)$, $f(h)$ and $\gamma(x)$ at zero. Moreover, $G(x) \in SV_{\infty}$, $\hat{m}(h) \in RV_0$ and $\hat{f}(h) \in RV_0$.

For $F \in DN(\alpha)$, relation (2.64) imply that

$$\log P\left(S_n \geq x_n n^{1/\alpha}\right) \sim -\lambda x_n^{1/\lambda} \tag{2.65}$$

for every sequence $\{x_n\}$ with $x_n \to \infty$ and $x_n = o(n^{\lambda})$. For $\alpha = 2$, this coincides with the result of Theorem 2.3.

For $F \in D(\alpha)$, the cases $\alpha < 2$ and $\alpha = 2$ are quite different. This is the result of a difference of conditions necessary and sufficient for attraction to the normal law and the stable laws with index $\alpha < 2$. The function $G(x)$ is the truncated second moment for $\alpha = 2$ and $G(x) \to \infty$ as $x \to \infty$. For $\alpha < 2$, it is the slowly varying part of the left-hand tail of the distribution of X and $G(x) \to \infty$ or 0 as $x \to \infty$.

The proof of Theorem 2.9 is similar to that of Theorem 2.6. We only need the next analogue of Lemma 2.21.

Lemma 2.22. *If* $\hat{m}(h)$, $\hat{f}(h)$ *and* $\hat{\gamma}(x)$ *are defined by formulae (2.62), then there exists a function* $L(x) \in SV_{\infty}$ *such that*

$$\hat{f}^{-1}(x) = \left(\frac{x}{c_4}\right)^{1/\alpha} L\left(\frac{1}{x}\right), \tag{2.66}$$

$$\hat{\gamma}(x) \sim \frac{c_2 x^{\lambda}}{c_4^{\lambda} L(1/x)} \quad as \quad x \to 0. \tag{2.67}$$

For $F \in DN(\alpha)$, we may directly use formulae (2.63) instead of Lemma 2.22 and calculations are simpler then.

The proof of Lemma 2.22 repeats that of Lemma 2.21 and therefore we omit details.

Note that $L(x) \to 0$ or ∞ in Lemma 2.22 while it only tends to ∞ for an attraction to the normal law. Moreover, $L(x)$ depends on the tail of the distribution of X. For $D(2)$, it is determined by the truncated second moment.

Now the proof of Theorem 2.9 repeats that of Theorem 2.6. We only mention that $\hat{\gamma}(cx_n) \sim c^\lambda \hat{\gamma}(x_n)$ for every fixed c. Changes of the calculations are obvious. Hence, we omit details.

We now turn to a result under the Linnik condition.

Theorem 2.10. *Assume that* $E\mathrm{e}^{(X^+)^\beta} < \infty$ *for some* $\beta \in (0,1)$, $EX = 0$ *and* $\alpha \in (1,2)$. *For* $F \in D(\alpha)$, *define* $\hat{m}(h)$, $\hat{f}(h)$ *and* $\hat{\gamma}(x)$ *by relation (2.62). For* $F \in DN(\alpha)$, *define* $\hat{m}(h)$, $\hat{f}(h)$ *and* $\hat{\gamma}(x)$ *by (2.63). Then relation (2.64) holds for every sequence* $\{x_n\}$ *with* $nx_n \to \infty$, $nx_n\left(L\left(1/x_n\right)\right)^{-1/\lambda} \to \infty$ *and* $n^{1-\beta}x_n^{1-\beta\lambda}\left(L\left(1/x_n\right)\right)^\beta \to 0$.

Note that the second condition on nx_n in Theorem 2.10 disappears provided $L(x) \to 0$ or c as $x \to \infty$. The latter holds for $F \in DN(\alpha)$. Remember that $L(x) \to 0$ as $x \to \infty$ for $F \in D(2)$. This was used in the proof of Theorem 2.7. When $F \in D(\alpha)$, $\alpha < 2$, we may have $L(x) \to \infty$ as $x \to \infty$ which gives a stronger condition on nx_n.

Theorem 2.10 implies that for $F \in DN(\alpha)$, relation (2.65) holds for every sequence $\{x_n\}$ with $x_n \to \infty$ and $x_n = o\left(n^{\beta\lambda(1-\lambda)/(1-\lambda\beta)}\right)$. For $\alpha = 2$, this corresponds to Theorem 2.4.

The proof of Theorem 2.10 repeats that of Theorem 2.7. We need only apply Lemma 2.22 instead of Lemma 2.21 and $\hat{\gamma}(cx_n) \sim c^\lambda \hat{\gamma}(x_n)$ for every fixed c. One can obviously change the calculations. Details are omitted.

We finally discuss the case of power moment assumptions.

Theorem 2.11. *Assume that* $E(X^+)^p < \infty$ *for some* $p > \alpha$, $EX = 0$ *and* $\alpha \in (1,2)$. *For* $F \in D(\alpha)$, *define* $\hat{m}(h)$, $\hat{f}(h)$ *and* $\hat{\gamma}(x)$ *by relation (2.62). For* $F \in DN(\alpha)$, *define* $\hat{m}(h)$, $\hat{f}(h)$ *and* $\hat{\gamma}(x)$ *by (2.63). Then relation (2.64) holds for every sequence* $\{x_n\}$ *with* $nx_n(\max\{1, \log L(1/x_n)\})^{-1} \to \infty$ *and* $\limsup(\alpha n x_n)/\log n \le p - \alpha$.

Note that the first condition on nx_n in Theorem 2.11 turns to $nx_n \to \infty$ when $L(x) \to 0$ or c as $x \to \infty$. The latter holds provided $F \in DN(\alpha)$. Remember that $L(x) \to 0$ as $x \to \infty$ for $F \in D(2)$. When $F \in D(\alpha)$, $\alpha < 2$, we may also have $L(x) \to \infty$ as $x \to \infty$ which yields a stronger condition on the rate of the growth of nx_n.

If $F \in DN(\alpha)$, then by Theorem 2.11, relation (2.65) holds for for every sequence $\{x_n\}$ with $x_n \to \infty$ and $\limsup x_n^{1/\lambda}/\log n \le (p-\alpha)/(\alpha-1)$. For $\alpha = 2$, this is the result of Theorem 2.5.

The proof of Theorem 2.11 is the same as that of Theorem 2.8. We need only apply Lemma 2.22 instead of Lemma 2.21 and $\hat{\gamma}(cx_n) \sim c^\lambda \hat{\gamma}(x_n)$ for every fixed c. Calculations may be changed obviously. We omit details.

2.13 Large Deviations and the Classification of Distributions

In this section, we give generalizations of the results of Section 2.2 related to the classification of distributions from Section 2.4.

Lemma 2.23. *Assume that X is non-degenerate, $EX \geq 0$ and $h_0 > 0$. Put $h_0' = \infty$ for $F \in K_5$ and $h_0' = h_0$ otherwise. Then*

$$P(S_n \geq nm(h)) \leq \mathrm{e}^{-nf(h)} \tag{2.68}$$

for all $h \in (0, h_0')$ and n.

Lemma 2.23 is a slight generalization of Lemma 2.2 for $F \in K_5$. It follows from the Tchebyshev inequality and the definitions of $m(h)$ and $f(h)$.

Lemma 2.24. *Assume that X is non-degenerate, $EX \geq 0$ and $h_0 > 0$. Let $\{h_n\}$ be a sequence of positive numbers. Assume that one of the following condition holds:*

1) $h_n < h_0$, $nf(h_n) \to \infty$ and $h_n = O(1)$;

2) $F \in K_1 \cup K_2$ and $h_n \to \infty$;

3) $F \in K_3 \cup K_5$, $h_n \to \infty$ and

$$P(X \geq (1-\tau)m(h)) \geq \mathrm{e}^{-(1+\delta)f(h)} \tag{2.69}$$

for all $\tau > 0$, $\delta > 0$ and all sufficiently large h;

4) $F \in K_4$, $h_n \nearrow h_0$ and inequality (2.69) holds for all $\tau > 0$, $\delta > 0$ and all h sufficiently close to h_0;

5) $F \in K_5$ and $h_n = h_* > h_0$.

Then

$$P(S_n \geq (1-\varepsilon)nm(h_n)) \geq \mathrm{e}^{-nf(h_n)(1+\delta)} \tag{2.70}$$

for all $\varepsilon > 0, \delta > 0$ and all sufficiently large n.

Proof. We first prove an auxiliary result.

Lemma 2.25. *If condition 1) holds, then $h_n\sigma(h_n) = o(\sqrt{n}f(h_n))$.*

Proof. If $0 < \varepsilon \le h_n \le 1/\varepsilon$ for all large n, then

$$h_n \sigma(h_n) \le \frac{1}{\varepsilon} \sup_{\varepsilon \le h \le 1/\varepsilon} \sigma(h) = O(1)$$

and $\sqrt{n} f(h_n) \ge \sqrt{n} f(\varepsilon) \to \infty$. Thus, we examine the case $h_n \to 0$. For $EX^2 < \infty$, the result follows by Lemma 2.7.

Assume that $EX^2 = \infty$. For $h > 0$, put $Y_h = \min\{h^2X^2, 1\}$. Then

$$EY_h = h^2 \int_{-1/h}^{1/h} x^2 dF(x) + P(|X| \ge 1/h).$$

It is clear that $h^2 = o(EY_h)$ as $h \to 0$.

Applying the inequality $\log x \le x - 1$ for $x \ge 1$, we have

$$f(h) \ge hm(h) - \varphi(h) + 1 = 1 - \varphi(h) + h\varphi'(h) - hm(h)(\varphi(h) - 1).$$

Taking into account that $1 - \mathrm{e}^v + v\mathrm{e}^v \ge (1 - 2\mathrm{e}^{-1}) \min\{v^2, 1\}$ for all real v, we have

$$1 - \varphi(h) + h\varphi'(h) = E(1 - \mathrm{e}^{hX} + hX\mathrm{e}^{hX}) \ge (1 - 2\mathrm{e}^{-1})EY_h.$$

Further, for $h \le \varepsilon_0 = \min\{h_0/2, 1\}$, we get

$$\varphi(h) - 1 \le \int_0^\infty (\mathrm{e}^{hx} - 1) dF(x) \le \int_0^\infty hx\mathrm{e}^{hx} dF(x) \le h \int_0^\infty x\mathrm{e}^{\varepsilon_0 x} dF(x).$$

It follows that $\varphi(h) - 1 = O(h)$ as $h \to 0$. Moreover,

$$hm(h)\varphi(h) = h\varphi'(h) \le h \int_0^\infty x\mathrm{e}^{hx} dF(x)$$

which yields that $hm(h) = O(h)$ in view of $\varphi(h) \to 1$ as $h \to 0$. Hence, $hm(h)(\varphi(h) - 1) = O(h^2) = o(EY_h)$ as $h \to 0$. Finally, we get

$$f(h) \ge \frac{1}{2}(1 - 2\mathrm{e}^{-1})EY_h$$

for all small h.

From the other hand, in view of $v^2\mathrm{e}^v \le 4\mathrm{e}^2 \min\{v^2, 1\}$ for $v \le 1$, we have

$$h^2\sigma^2(h)\varphi(h) \le \int_{-\infty}^\infty h^2x^2\mathrm{e}^{hx} dF(x) \le 4\mathrm{e}^2 EY_h + h^2 \int_{-\infty}^\infty x^2\mathrm{e}^{\varepsilon_0 x} dF(x)$$

provided $h \le \varepsilon_0$. It follows that

$$h^2\sigma^2(h) \le 5\mathrm{e}^2 EY_h \le 10\mathrm{e}^2(1 - 2\mathrm{e}^{-1})^{-1} f(h)$$

for all small h. This yields the result. □

If condition 1) holds and $EX = 0$ then inequality (2.70) follows from Lemmas 2.2 and 2.25. If $EX > 0$, then

$$P(S_n \geq (1-\varepsilon)nm(h_n)) \geq P(S_n - nEX \geq (1-\varepsilon)n(m(h_n) - EX)).$$

Note that a shift of X does not change $f(h)$ while $m(h)$ increases on the value of the shift. Hence, the result follows.

Suppose that condition 2) is satisfied. Then

$$P(S_n \geq (1-\varepsilon)nm(h_n)) \geq (P(X \geq (1-\varepsilon)\omega))^n = \mathrm{e}^{-n\log P(X\geq(1-\varepsilon)\omega)}.$$

If $F \in K_1$, then this bound and $f(h_n) \to \infty$ yield inequality (2.70). If $F \in K_2$, then we apply the same bound with $\varepsilon = 0$ and $f(h_n) = -\log P(X = \omega)(1+o(1))$.

If either condition 3) or condition 4) holds, then (2.70) follows from (2.69) and the inequality $P(S_n \geq nx) \geq (P(X \geq x))^n$.

Assume that condition 5) is satisfied.

We first check that for all $\tau \in (0,1)$, $\delta > 0$ and $H > 0$ there exists $h > H$ such that

$$P(h \geq X \geq (1-\tau)h) \geq \mathrm{e}^{-(1+\delta)h_0 h}. \tag{2.71}$$

Suppose that the latter fails. Then there exist $\tau \in (0,1)$, $\delta > 0$ and $H > 0$ such that for all $h > H$, the inequality opposite to (2.71) holds. Put $t_k = (1-\tau)^{-k}$, $k \in \mathbf{N}$. For all sufficiently large k, the inequality

$$\int_{(1-\tau)t_k}^{t_k} \mathrm{e}^{hx} dF(x) \leq \mathrm{e}^{ht_k} P((1-\tau)t_k \leq X \leq t_k) < \mathrm{e}^{(h-(1+\delta)h_0)t_k}$$

holds. Hence, $\varphi(h) < \infty$ for $0 < h < (1+\delta)h_0$ which contradicts to the definition of h_0. Inequality (2.71) is proved.

Let $\bar{X}, \bar{X}_1, \bar{X}_2, \ldots$ be a sequence of independent random variables with the d.f.

$$\frac{1}{\varphi(h_0)} \int_{-\infty}^{x} \mathrm{e}^{h_0 t} dF(t).$$

Denote $\bar{S}_n = \bar{X}_1 + \cdots + \bar{X}_n$ and $\bar{F}_n(t) = P(\bar{S}_n < t)$.

Fix a positive number δ and put $\tau = \delta/2$, $\tau = 6\rho$.

Put $z = h_* - h_0$, $x = nm(h_*)$, $y = x + \tau nz$, $Q_n = P(x \leq \bar{S}_n \leq y)$.

By the definition of the conjugate distribution, we have

$$P_n = P(S_n \geq x) = (\varphi(h_0))^n \int_x^{\infty} \mathrm{e}^{-th_0} d\bar{F}_n(t) \geq (\varphi(h_0))^n \mathrm{e}^{-h_0 y} Q_n. \tag{2.72}$$

Let $a > 1$ be a fixed number which will be chosen later. Put $T_n = \bar{S}_n - E\bar{S}_n$, $u = m(h_0) + (1+2\rho)za$, $v = m(h_0) + (1+3\rho)za$. Taking into account (2.5) and $E\bar{S}_n = nm(h_0)$, we have

$$\begin{aligned} Q_n &= P(zn \le T_n \le (1+6\rho)zn) \\ &\ge P((1+2\rho)zn \le T_{n/a} \le (1+3\rho)zn)P(-2\rho zn \le T_n - T_{n/a} \le 3\rho zn) \\ &\ge \frac{1}{2}P((1+2\rho)zn \le T_{n/a} \le (1+3\rho)zn) \ge \frac{1}{2}(P(u \le \bar{X} \le v))^{n/a} \end{aligned}$$

for all sufficiently large n by the weak law of large numbers. We get

$$Q_n \ge \frac{1}{2}\Big(\frac{1}{\varphi(h_0)}\int_u^v \mathrm{e}^{h_0 t}dF(t)\Big)^{n/a} \ge \frac{1}{2}\Big(\frac{e^{h_0 u}}{\varphi(h_0)}P(u \le X \le v)\Big)^{n/a} \tag{2.73}$$

for all sufficiently large n. Note that $\lim_{a\to\infty} u/v = \varepsilon \in (0,1)$. So, if a is large enough, then

$$P(u \le X \le v) \ge P(v\varepsilon^{1/2} \le X \le v) \ge \exp\Big\{-\frac{1+4\rho}{1+3\rho}h_0 v\Big\}. \tag{2.74}$$

In the last inequality, we have applied (2.71). Relations (2.72)–(2.74) imply that

$$P_n \ge \frac{1}{2}\mathrm{e}^{np(a)} \tag{2.75}$$

for all sufficiently large n, where

$$\begin{aligned} p(a) &= \log\varphi(h_0) - \frac{h_0 y}{n} + \frac{h_0 u}{a} - \frac{1}{a}\log\varphi(h_0) - \frac{(1+4\varrho)h_0 v}{(1+3\varrho)a} \\ &= \log\varphi(h_0) - h_0(m(h_*) + 6\varrho z) + \frac{h_0}{a}(m(h_0) + (1+2\varrho)za) \\ &\quad -\frac{1}{a}\log\varphi(h_0) - \frac{(1+4\varrho)h_0}{(1+3\varrho)a}(m(h_0) + (1+3\varrho)za). \end{aligned}$$

In the last equality, we have used the definitions of y, u, v and $\tau = 6\varrho$. It yields that $p(a)$ does not depend on n and

$$\begin{aligned} &\lim_{a\to\infty} p(a) = \log\varphi(h_0) - h_0 m(h_*) - 6h_0\varrho z + (1+2\varrho)h_0 z - (1+4\varrho)h_0 z \\ &= -f(h_*) - 8\rho h_0 z \ge -(1+8\rho)f(h_*). \end{aligned}$$

We have used (2.5) in the last equality. If a is large enough, then (2.75) implies (2.70). □

2.14 Bibliographical Notes

An asymptotic behaviour of probabilities of large deviations has been first investigated in [Khintchine (1929a,b)]. General results for large deviations have been proved in [Cramér (1938)]. Generalizations of these results was proved by [Feller (1943)] and [Petrov (1954)]. Under the Cramér condition, limit behaviour of probabilities of large deviations for independent, non-identically distributed random variables has been studied by [Petrov (1954, 1961, 1968)], [Feller (1943)], [Statulevičius (1966)] and many others. Results in this topic may be found in monographs [Petrov (1975, 1995)], for example. A generalization of results from [Petrov (1954)] for a scheme of series has been obtained in [Frolov *et al.* (1997)].

[Linnik (1961a,b,c, 1962)] has developed methods of investigations of asymptotic behaviour for large deviations probability when the Cramér conditions is violated. Results and the techniques may be found in [Ibragimov and Linnik (1971)]. Linnik's researches was extended by [Petrov (1963, 1964)], [Nagaev (1965)], [Wolf (1968, 1970)], [Osipov (1972)] and others.

Various one-sided results on large deviations have been obtained by [Chernoff (1952)], [Daniels (1954)], [Bahadur and Ranga Rao (1960)], [Zolotarev (1962)], [Borovkov (1964)], [Petrov (1965)], [Feller (1969)], [Osipov (1972)], [Rozovskii (1999, 2001, 2003)], [Nagaev (1969)], [Frolov (1998, 2002c)] and other.

Results for probabilities of moderate deviations have been proved by [Amosova (1972, 1979, 1980)], [Rozovskii (1981, 2004)], [Slastnikov (1978, 1984)], [Michel (1976)], [Rychlik (1983)], [Frolov (1998, 2002b, 2008)] and references therein.

Results for large deviations in the case of attraction to stable laws may be found in [Lipschutz (1956b)], [Bahadur and Ranga Rao (1960)], [Kim and Nagaev (1975)], [Höglund (1979)], [Nagaev (1981)], [Nagaev (1983)], [Amosova (1984)], [Rozovskii (1989a,b, 1993, 1997)] and references therein.

Bibliography on related topics (large deviations on the hole line, local limit theorems in relation to large deviations etc.) may be found in the book [Petrov (1975)]. Large deviations of sums of independent random variables are also discussed in monograph [Borovkov and Borovkov (2008)].

Note that we do not discuss large deviation principles in this book.

Properties of stable laws are considered in [Mijnheer (1974)] and [Samorodnitsky and Taqqu (1994)].

Lemma 2.2 is proved by [Feller (1969)]. Theorem 2.3 follows from the main result of the last paper. Theorem 2.2 is proved by [Chernoff (1952)]. In the Chapter 2, we follows a pattern of the papers [Frolov (1998, 2002c, 2003b)]. Partially, results of Chapter 2 are discussed in [Frolov (2014)].

Chapter 3

Strong Limit Theorems for Sums of Independent Random Variables

Abstract. A universal theory of strong limit theorems is discussed. Starting with a formula for norming sequences, we prove universal strong laws. Then we show that they imply the Erdős–Rényi law and its Mason's extension, the Shepp law, the Csörgő–Révész laws, the SLLN and the LIL. Moment assumptions are either optimal, or close to optimal.

3.1 Norming Sequences in Strong Limit Theorems

Let $X, X_1, X_2, \ldots$ be a sequence of i.i.d. random variables with a non-degenerate d.f. $F(x) = P(X < x)$ and $EX \geq 0$. Put

$$S_n = X_1 + X_2 + \cdots + X_n, \quad S_0 = 0.$$

Assume that $a(x)$ is a non-decreasing, continuous function such that $1 \leq a(x) \leq x$ and $x/a(x)$ is non-decreasing. Put $a_n = [a(n)]$, where $[\cdot]$ is the integer part of the number in brackets. Denote

$$U_n = \max_{0 \leq k \leq n - a_n} (S_{k+a_n} - S_k),$$

$$W_n = \max_{0 \leq k \leq n - a_n} \max_{1 \leq j \leq a_n} (S_{k+j} - S_k),$$

$$R_n = S_n - S_{n-a_n}, \quad T_n = S_{n+a_n} - S_n.$$

In this section, we give a formula for norming sequences in strong limit theorems for increments, i.e. a formula for $\{b_n\}$ such that

$$\limsup \frac{W_n}{b_n} = 1 \quad \text{a.s.}$$

Put $h_0 = \sup\{h : Ee^{hX} < \infty\}$.

Assume first that $h_0 = 0$. It follows that $\omega = \infty$. Let $\{y_n\}$ be a sequence of positive numbers such that $y_n \to \infty$. For $n \in \mathbf{N}$, put

$$Y_n = \min\{X, y_n\}, \tag{3.1}$$

and

$$Z_n = \begin{cases} Y_n - EY_n, & \text{for} \quad EX = 0, \\ Y_n, & \text{for} \quad EX > 0. \end{cases}$$

Random variables Z_n are bounded from above. Hence, they have an exponential moment.

For every Z_n, $n \in \mathbf{N}$, define the functions $\varphi_n(h)$, $m_n(h)$, $\sigma_n^2(h)$, $f_n(h)$, $\zeta_n(z)$ and $\gamma_n(x)$ by

$$\begin{aligned} &\varphi_n(h) = Ee^{hZ_n}, \quad m_n(h) = \frac{\varphi_n'(h)}{\varphi_n(h)}, \\ &\sigma_n^2(h) = m_n'(h), \quad f_n(h) = hm_n(h) - \log \varphi_n(h), \\ &\zeta_n(z) = \sup\{zh - \log \varphi_n(h) : h \geq 0, \varphi_n(h) < \infty\}, \\ &\gamma_n(x) = \sup\{z : \zeta_n(z) \leq x\}. \end{aligned}$$

Assume now that $h_0 > 0$. For $n \in \mathbf{N}$, put $y_n = \infty$ in (3.1) and define $\varphi_n(h)$, $m_n(h)$, $\sigma_n^2(h)$, $f_n(h)$, $\zeta_n(z)$ and $\gamma_n(x)$ by the above formulae. Of course, we assume that $h < h_0$ and x and z are from corresponding regions in there.

It is clear that for $h_0 > 0$, we have $Z_n = Y_n = X$, $\varphi_n(h) = \varphi(h)$, $m_n(h) = m(h)$, $\sigma_n^2(h) = \sigma^2(h)$, $f_n(h) = f(h)$, $\zeta_n(z) = \zeta(z)$ and $\gamma_n(x) = \gamma(x)$ for all n, where $\varphi(h)$, $m(h)$, $\sigma^2(h)$, $f(h)$, $\zeta(z)$ and $\gamma(x)$ are the functions of the LDT discussed in Chapter 2.

We now write the formula for $\{b_n\}$. For $n \in \mathbf{N}$, put

$$b_n = a_n \gamma_n(d_n), \tag{3.2}$$

where

$$d_n = \frac{\beta_n}{a_n} \quad \text{and} \quad \beta_n = \log \frac{n}{a_n} + \log\log(\max(n, 3)). \tag{3.3}$$

Formula (3.2) is one of main results of this book. We will show that relation (3.2) defines norming sequences in the SLLN, the Erdős–Rényi law and its Mason's extension, the Shepp law, the Csörgő–Révész laws and the LIL.

Relation (3.2) allows to calculate $\{b_n\}$ for all $\{a_n\}$ provided $h_0 > 0$. If $h_0 = 0$, then to find $\{b_n\}$ we need an asymptotic of $\gamma_n(d_n)$ which crucially depends on a choice of $\{y_n\}$. This can not be done without additional assumptions on the distribution of X. We will check in the same way as in Chapter 2 that for distributions from the domains of attraction of the normal law and the asymmetric stable laws with index $\alpha \in (1, 2)$, one can choose $\{y_n\}$ such that

$$\gamma_n(d_n) = \hat{\gamma}(d_n)(1 + o(1)), \tag{3.4}$$

where $\hat{\gamma}(x) = C_\alpha x^{(\alpha-1)/\alpha} L(x)$ and $L(x) \in SV_0$. Note that relation (3.4) only holds for $a_n/\log n \to \infty$. Indeed, if $a_n = c \log n$ and (3.4) holds, then $b_n \sim c_1 \log n$. Hence, the condition $\sum_n P(X \geq b_n) < \infty$ implies $h_0 > 0$.

Assume that either $Ee^{(X^+)^\beta} < \infty$, $\beta \in (0,1)$, or $E(X^+)^p < \infty$, $p > 2$.

We know from Chapter 2 that under some restriction on $\{y_n\}$ and $\{h_n\}$ with $y_n \to \infty$ and $h_n \to 0$, the relations

$$m_n(h_n) = \hat{m}(h_n)(1 + o(1))$$

and

$$f_n(h_n) = \hat{f}(h_n)(1 + o(1))$$

hold, where $\hat{m}(h)$ and $\hat{f}(h)$ are positive, strictly increasing functions. Put $h_n = \hat{f}^{-1}(d_n)$ and $\hat{\gamma}(x) = \hat{m}(\hat{f}^{-1}(x))$. Take $y_n = a_n \hat{\gamma}(d_n)$, if the Linnik condition is satisfied, and $y_n = 1/h_n$, otherwise. Then (3.4) holds. Moreover, relations between $\{y_n\}$ and $\{h_n\}$ yield restrictions on the minimal rate of the growth of a_n for which (3.4) holds.

For example, if $EX = 0$ and $EX^2 = 1$, then $\hat{m}(h) = h$, $\hat{f}(h) = h^2/2$, $\hat{\gamma}(x) = \sqrt{2x}$. It follows that $b_n = \sqrt{2a_n\beta_n}$. When the Linnik condition is fulfilled, this formula holds for $a_n \geq C_\beta(\log n)^{2/\beta - 1}$, where C_β is a certain positive constant. If $E(X^+)^p < \infty$, $p > 2$, then $a_n \geq Cn^{2/p}/\log n$ has to be satisfied, where C is an arbitrary positive constant.

If $h_0 > 0$ and $a_n/\log n \to \infty$, then we need an asymptotic of $\gamma(x)$ at zero to find a simple formula for b_n. This problem may be solved in the same way as a deduction of (3.4). In particular, for $EX = 0$ and $EX^2 = 1$, we have $b_n = \sqrt{2a_n\beta_n}$ again. This corresponds to the limit case $\beta = 1$ in the Linnik condition.

Hence, formula (3.2) is applicable for $h_0 > 0$ and $h_0 = 0$ both. We will show below that (3.2) determines norming sequences in the SLLN, the Erdős–Rényi law and its Mason's extension, the Shepp law, the Csörgő–Révész laws and the LIL.

3.2 Universal Strong Laws in Case of Finite Exponential Moments

We assume in this section that the random variable X satisfies the Cramér condition. Since $h_0 > 0$, the functions $\varphi_n(h)$, $m_n(h)$, $\sigma_n^2(h)$, $f_n(h)$, $\zeta_n(z)$ and $\gamma_n(x)$ coincide for all n with $\varphi(h)$, $m(h)$, $\sigma^2(h)$, $f(h)$, $\zeta(z)$ and $\gamma(x)$ which are discussed in Chapter 2. In this case, formula (3.2) turns to

$$b_n = a_n \gamma(d_n)$$

for $n \in \mathbf{N}$ with d_n from (3.3).

The universal results of Section 1.1 contain conditions on distributions of sums S_n. Our goal is to derive results under restrictions on the distribution of X. It turns out that they crucially depend on classes of distributions from Section 2.4.

We start with upper bounds.

Theorem 3.1. *Assume that one of the following conditions holds:*

1) $\log a_n / \log n \to 0$;

2) *for every* $\varepsilon > 0$ *there exists* $q \in (0,1)$ *such that inequality (1.14) of Theorem 1.1 holds for all sufficiently large* n.

Then

$$\limsup \frac{W_n}{b_n} \leq 1 \quad a.s.$$

One can replace W_n *by* T_n *in the last relation.*

We need the following lemma to check property (1.12) of $\{b_n\}$.

Lemma 3.1. *Let* $g(x)$ *be a continuous, positive, non-decreasing, concave function with* $g(0) \geq 0$. *For all* $n \in \mathbf{N}$, *put* $b_n = a_n g(d_n)$. *Then there exists a non-decreasing sequence* $\{\tilde{b}_n\}$ *such that* $b_n \sim \tilde{b}_n$ *and (1.12) holds.*

Proof. Check first that for every $\tau \in (0,1)$, there exists N such that $\beta_{n+k} \geq (1-\tau)\beta_n$ for all $n \geq N$ and $k \in \mathbf{N}$.

If the function $a(x)$ is bounded, then $a_n = const$ and $\beta_{n+k} \geq \beta_n$ for all sufficient large n. Assume that $a(x) \to \infty$ as $x \to \infty$. Since $x/a(x)$ is non-decreasing, we have

$$\beta_{n+k} \geq \log\Big(\frac{n+k}{a(n+k)} \log n\Big) \geq \log\Big(\frac{n}{a(n)} \log n\Big) \sim \beta_n$$

for all fixed $k \in \mathbf{N}$.

Using the concavity of $g(x)$, we conclude that for every $\tau \in (0,1)$ there exists N such that

$$\frac{b_{n+k}}{b_n} = \frac{a_{n+k}}{a_n}\frac{g(d_{n+k})}{g(d_n)} \geq \frac{g(\frac{\beta_{n+k}}{a_n})}{g(d_n)} \geq \frac{g((1-\tau)d_n)}{g(d_n)} \geq 1-\tau \tag{3.5}$$

for all $n \geq N$ and $k \in \mathbf{N}$.

Put $\tilde{b}_n = \inf_{k \geq 0} b_{n+k}$. It is clear that $\tilde{b}_{n+1} \geq \tilde{b}_n$ and $\tilde{b}_n \leq b_n$. By (3.5), we get $\tilde{b}_n \sim b_n$.

For $k \in \mathbf{N}$, denote $n_k = [\theta^k]$, $\theta > 1$. The function $a(x)$ is non-decreasing and $x/y \geq \log x / \log y$ for $x \geq y \geq e$. It follows that

$$r_k = \frac{a_{n_k}\beta_{n_{k+1}}}{\beta_{n_k} a_{n_{k+1}}} \leq \frac{\beta_{n_{k+1}}}{\beta_{n_k}} \leq \frac{\log(\frac{n_{k+1}}{a_{n_k}} \log n_{k+1})}{\log(\frac{n_k}{a_{n_k}} \log n_k)} \leq \frac{n_{k+1} \log n_{k+1}}{n_k \log n_k} \leq \theta^2$$

for all sufficiently large k. The function $x/a(x)$ is non-decreasing and, therefore,

$$\frac{a_{n_{k+1}}}{a_{n_k}} \le \theta \frac{a(n_{k+1})}{a(n_k)} \le \theta \frac{n_{k+1}}{n_k} \le \theta^3$$

for all sufficiently large k provided $a(x) \to \infty$ as $x \to \infty$. If $a(x)$ is bounded then the latter evidently holds.

By the concavity of $g(x)$, we have

$$\frac{b_{n_{k+1}}}{b_{n_k}} = \frac{a_{n_{k+1}}}{a_{n_k}} \frac{g(r_k d_{n_k})}{g(d_{n_k})} \le \theta^3 \frac{g(\theta^2 d_{n_k})}{g(d_{n_k})} \le \theta^5$$

for all sufficiently large k. Choosing θ with $\theta^5 \le 1 + \tau$, we get (1.12) from the last relation and (3.5). □

We also need the next result.

Lemma 3.2. *For all $x \ge 0$ and $\delta > 0$, one has $\zeta((1+\delta)\gamma(x)) \ge (1+\delta)x$.*

Note that the left-hand side of the last inequality may be infinite.

Proof. If $(1+\delta)\gamma(x) < A$ (for $A = \infty$ and $A < \infty$ both), then we apply the inequality $(1+\delta)\gamma(x) \ge \gamma((1+\delta)x)$ which follows from the concavity of $\gamma(x)$ and $\gamma(0) = EX \ge 0$. We have

$$\zeta((1+\delta)\gamma(x)) \ge \zeta(\gamma((1+\delta)x)) = (1+\delta)x.$$

For $(1+\delta)\gamma(x) > A$, we have to consider the classes K_1, K_2 and K_5. If $F \in K_1 \cup K_2$, then $(1+\delta)\gamma(x) > \omega$ that yields $\zeta((1+\delta)\gamma(x)) = \infty$. For $F \in K_5$, we get

$$\zeta((1+\delta)\gamma(x)) = (1+\delta)\gamma(x)h_0 - \log \varphi(h_0) \ge (1+\delta)x$$

in view of $\gamma(x) \ge (x + \log \varphi(h_0))/h_0$ for all $x \ge 0$. The latter follows from the concavity of $\gamma(x)$ and $\gamma'(1/c_0) = 1/h_0$. □

We now turn to the proof of Theorem 3.1.

Proof. By Lemma 3.1, relation (1.12) holds.

Assume that condition 2) holds. Making use of Lemma 3.2 and the definitions of $\gamma(x)$ and $\zeta(z)$, we have

$$P(S_{[(1+\varepsilon)a_n]} \ge (1+\varepsilon)b_n) \le P(S_{[(1+\varepsilon)a_n]} \ge [(1+\varepsilon)a_n]\gamma(d_n))$$
$$\le \mathrm{e}^{-[(1+\varepsilon)a_n]\beta_n/a_n}.$$

This yields (1.13) with $H = 0$.

Suppose that condition 1) holds. By the definition of $\zeta(z)$ and Lemma 3.2, we have

$$P(S_k \geq (1+\varepsilon)b_n) \leq \mathrm{e}^{-k\zeta((1+\varepsilon)b_n/k)} = \mathrm{e}^{-k\zeta((1+\varepsilon)a_n\gamma(d_n)/k)}$$
$$\leq \mathrm{e}^{-(1+\varepsilon)a_n d_n} = \mathrm{e}^{-(1+\varepsilon)\beta_n}$$

for all $k \leq [(1+\varepsilon)a_n]$. This implies (1.13) with $H=0$ for S_k.

Theorem 3.1 follows from Theorem 1.1. □

We turn now to lower bounds.

Theorem 3.2. *For every $\varepsilon \in (0,1)$, put $h^* = f^{-1}((1-\varepsilon)d_n)$. Assume that one of the following conditions holds:*

1) $h^* < h_0$ and $h^* = O(1)$;

2) $F \in K_1$ and $h^* \to \infty$;

3) $F \in K_2$ and $d_n \geq 1/c_0$;

4) $F \in K_3 \cup K_5$, $h^* \to \infty$ and inequality (2.69) holds for all $\tau > 0$, $\delta > 0$ and all sufficiently large h;

5) $F \in K_4$, $h^* \nearrow h_0$ and inequality (2.69) holds for all $\tau > 0$, $\delta > 0$ and all h sufficiently close to h_0;

6) $F \in K_5$, $h^* > h_0$, h^* does not depend on n.

If $a_n/n \to 1$, then assume in addition that condition 2) of Theorem 3.1 holds.

Then

$$\limsup \frac{R_n}{b_n} \geq 1 \quad a.s.$$

One can replace R_n by T_n in the last relation.

Note that h_0 may be infinite and $h^* = O(1)$ is a restriction in this cases.

Conditions 1)–6) of Theorem 3.2 allow to consider various sequences $\{a_n\}$ for all five classes of distributions. Condition $h^* = O(1)$ is equivalent to $\liminf a_n/\log n > 0$ while $h^* \to \infty$ is equivalent to $a_n = o(\log n)$.

If $F \in K_1$, then one has to check condition 1) for $\liminf a_n/\log n > 0$ and condition 2) for $a_n = o(\log n)$.

If $F \in K_2$, then the function $f(h)$ takes its values in $[0, 1/c_0)$. Hence, one can use condition 1) for $a_n > c_0 \log n$ only. The case $a_n \leq c_0 \log n$ is included in 3).

If $F \in K_3 \cup K_4$, then condition 1) may be applied for $\liminf a_n/\log n > 0$. Conditions 4) and 5) may be simpler for $a_n = o(\log n)$.

If $F \in K_5$, then one can use condition 1) for $\liminf a_n/\log n > 0$, condition 6) for a fixed $h^* > h_0$ ($a_n = c\log n, 0 < c < c_0$) and condition 4) for $a_n = o(\log n)$.

Note that condition 1) and one of conditions 2)–6) may be satisfied simultaneously.

Turn to the proof of Theorem 3.2.

Proof. We first check that relation (1.23) holds. Denote

$$A_n^\varepsilon = \{S_{a_n} \geq (1-\varepsilon)b_n\},$$

where $\varepsilon \in (0,1)$.

Suppose that one of conditions 1), 2), 4)–6) is satisfied. By the concavity of $\gamma(x)$, the inequality $(1-\varepsilon)^2 b_n \leq (1-\varepsilon)a_n\gamma((1-\varepsilon)d_n)$ holds. Using (2.70) and the definition of $\gamma(x)$, we have

$$\begin{aligned}
&P(A_n^{2\varepsilon-\varepsilon^2}) = P(S_{a_n} \geq (1-\varepsilon)^2 b_n) \\
&\geq P\left(S_{a_n} \geq (1-\varepsilon)a_n\gamma((1-\varepsilon)d_n)\right) \geq \mathrm{e}^{-(1+\tau)(1-\varepsilon)\beta_n}
\end{aligned}$$

for all sufficiently large n. Taking τ small enough, we arrive at (1.23).

If condition 3) is fulfilled, then $b_n = a_n\omega$. We assume further that $a_n \to \infty$ since (1.23) automatically holds otherwise. Let m be a natural number which will be chosen later. We have

$$P(A_n^\varepsilon) \geq (P(S_m \geq (1-\varepsilon)\omega m))^{a_n/m} \geq (P(S_m \geq (1-\varepsilon)\omega m))^{c_0\beta_n/m}.$$

Take x_0 with $F(x_0) < F(\omega)$. Chose m large enough to satisfy $(1-\varepsilon)m\omega \leq x_0 + (m-1)\omega$. Then

$$\begin{aligned}
&P(S_m \geq (1-\varepsilon)\omega m) \geq P(S_m \geq x_0 + (m-1)\omega) \\
&\geq mP(\omega > X_1 \geq x_0, X_2 = \cdots = X_m = \omega) \\
&= mP(\omega > X \geq x_0)(P(X=\omega))^{m-1} = m(F(\omega) - F(x_0))\mathrm{e}^{-(m-1)/c_0}.
\end{aligned}$$

This yields (1.23).

Theorem 3.2 now follows from Theorem 1.2 and Lemma 3.1. □

Our next result is a lower bound for U_n when $\{a_n\}$ increases slow enough.

Theorem 3.3. *If the conditions of Theorem 3.2 hold and* $\log\log n = o(\log(n/a_n))$, *then*

$$\liminf \frac{U_n}{b_n} \geq 1 \quad a.s.$$

Proof. We checked in the proof of Theorem 3.2 that relations (1.23) and (1.12) hold. Theorem 3.3 now follows from Theorem 1.3. □

Theorems 3.1–3.3 and the inequality $R_n \le U_n \le W_n$ imply the following result.

Theorem 3.4. *If the conditions of Theorems 3.1 and 3.2 hold, then*

$$\limsup \frac{W_n}{b_n} = \limsup \frac{U_n}{b_n} = \limsup \frac{R_n}{b_n} = \limsup \frac{T_n}{b_n} = 1 \quad a.s.$$

If $\log\log n = o(\log(n/a_n))$ *in addition, then*

$$\lim \frac{W_n}{b_n} = \lim \frac{U_n}{b_n} = 1 \quad a.s.$$

3.3 Universal Strong Laws for Random Variables without Exponential Moment

In this section, we investigate the asymptotic behaviour of increments of sums of i.i.d. random variables when the Cramér condition is violated. Then $h_0 = 0$ and we define the norming sequence $\{b_n\}$ by formula (3.2) in which $\gamma(x)$ is replaced by a similar function constructed from truncated random variables.

We first deal with upper bounds again.

Theorem 3.5. *Assume that the sequence* $\{b_n\}$ *is equivalent to a non-decreasing sequence and condition (1.12) of Theorem 1.1 holds.*

Assume that one of the following conditions holds:

1) $\sum_n P(X \ge b_n) < \infty$ *and* $y_n = b_n$.

2) $\sum_n P(X \ge cb_n) < \infty$ *for every* $c > 0$, *and for all small enough* $\varepsilon > 0$ *there exists* $\delta > 0$ *such that*

$$a_n P(X \ge y_n) \mathrm{e}^{-((1+\varepsilon)a_n - 1)\zeta_n((1-\delta)\gamma_n(d_n))} + (a_n P(X \ge y_n))^2 \le \mathrm{e}^{-(1+\varepsilon)\beta_n} \quad (3.6)$$

for all sufficiently large n.

Suppose that for every $\varepsilon > 0$ *there exists* $q \in (0,1)$ *such that inequality (1.14) of Theorem 1.1 holds for all sufficiently large* n.

Then

$$\limsup \frac{W_n}{b_n} \le 1 \quad a.s.$$

One can replace W_n *by* T_n *in the last relation.*

Proof. Assume that condition 1) holds. Define n_k by formula (1.17) with $c = 1$ from the proof of Theorem 1.1. Then the series $\sum_k n_k P(X \ge b_{n_k})$ converges. This implies that $a_{n_k} P(X \ge b_{n_k}) \to 0$ as $k \to \infty$. By Lemma

2.4 with $k = [(1+\varepsilon)a_{n_k}]$, $y = y_{n_k}$ and $x = \gamma_{n_k}(d_{n_k})$, we get (1.22) for all sufficiently large k. We obtain the result by Remark 1.2.

Assume that condition 2) holds. Since $\beta_n \to \infty$, relation (3.6) yields that $a_n P(X \geq y_n) \to 0$. By Lemma 2.4 with $k = [(1+\varepsilon)a_n]$, $y = y_n$ and $x = \gamma_n(d_n)$, we get (1.13) for all sufficiently large n. The result follows by Theorem 1.1. □

Turn now to lower bounds.

Theorem 3.6. *Assume that one of the following conditions holds:*

1) $\liminf \gamma_n(d_n) > 0$, *and for every* $\varepsilon > 0$ *there exists* $\tau > 0$ *such that*

$$a_n P(X \geq (1-\varepsilon)b_n) \geq \mathrm{e}^{-(1-\tau)\beta_n}$$

for all sufficiently large n.

2) *The inequality*

$$d_n \leq -\log P(X \geq y_n) \tag{3.7}$$

holds for all sufficiently large n *and*

$$h^* \sigma_n(h^*) = o(\sqrt{a_n} f_n(h^*)), \tag{3.8}$$

where $h^* = f_n^{-1}((1-\varepsilon)d_n)$, $\varepsilon \in (0,1)$, *and for* $EX = 0$,

$$EY_n = o(\gamma_n(d_n)). \tag{3.9}$$

If $a_n/n \to 1$, *then suppose in addition that (1.14) and (1.12) hold. Then*

$$\limsup \frac{R_n}{b_n} \geq 1 \quad a.s.$$

One can replace R_n *by* T_n *in the last relation.*

Note that the distribution of Y_n is from the class K_2 and the function $f_n(h)$ is bounded from above by $-\log P(X \geq y_n)$. Hence, h^* is well defined for all n such that (3.7) is satisfied.

Proof. Check that inequality (1.23) holds for all sufficiently large n.

Suppose that condition 1) is satisfied.

Assume first that $a_n = C_1$ for all sufficiently large n and $\limsup \gamma_n(d_n) < \infty$. Then $b_n \leq C_2$ and

$$P(S_{a_n} \geq (1-\varepsilon)b_n) \geq P(S_{a_n} \geq (1-\varepsilon)C_2) = p$$

for all sufficiently large n. If $p = 0$, then

$$0 = P(S_{a_n} \geq (1-\varepsilon)C_2) \geq \left(P\left(X \geq (1-\varepsilon)\frac{C_2}{C_1}\right)\right)^{C_1}.$$

It follows that the random variable X is bounded from above. This contradicts to the equality $h_0 = 0$. It yields that $p > 0$. Hence, the left-hand side of inequality (1.23) is separated from zero and the right-hand one tends to zero since $\beta_n \to \infty$. It implies that inequality (1.23) holds for all sufficiently large n.

Assume now that $a_n = C_1$ for all sufficiently large n and $\limsup \gamma_n(d_n) = \infty$. Then there are two sequences $\{n'_m\}$ and $\{n''_m\}$ such that $\{n'_m\} \bigcup \{n''_m\} = \mathbf{N}$, $b_{n'_m} \to \infty$ as $m \to \infty$ and $b_{n''_m} \le C_3$ for all sufficiently large m. For $n \in \{n''_m\}$, we prove in the same way as above that

$$P(S_{a_n} \ge (1-\varepsilon)b_n) \ge p > 0$$

and, therefore, (1.23) holds for all sufficiently large $n \in \{n''_m\}$. Consider now $n \in \{n'_m\}$. We have

$$(a_n - 1)P(X \ge (1-\varepsilon^2)b_n) \le \frac{a_n - 1}{(1-\varepsilon^2)b_n} EX^+ \to 0.$$

Since $EX \ge 0$ and X is non-degenerate, we have

$$P(S_{a_n} \ge -\varepsilon(1-\varepsilon)b_n) \ge P(S_{a_n} \ge 0) \ge (P(X \ge 0))^{a_n} = C_4 > 0.$$

Hence,

$$P(S_{a_n} \ge -\varepsilon(1-\varepsilon)b_n) - (a_n - 1)P(X \ge (1-\varepsilon^2)b_n) \ge \frac{C_4}{2}$$

for all sufficiently large $n \in \{n'_m\}$. By Lemma 2.6 with $\delta = \varepsilon$, $x = (1-\varepsilon)\gamma_n(d_n)$ and $k = a_n$, inequality (1.23) holds for all sufficiently large $n \in \{n'_m\}$ as well.

Assume now that $a_n \to \infty$. Since $\gamma_n(d_n) > \varrho > 0$ for all sufficiently large n, we have

$$P(S_{a_n} \ge -\varepsilon b_n) \ge P(S_{a_n} \ge -\varepsilon \varrho a_n) \to 1$$

by the weak law of large numbers. Moreover,

$$(a_n - 1)P(X \ge (1-\varepsilon^2)b_n) \le \frac{a_n - 1}{(1-\varepsilon^2)b_n} \int\limits_{(1-\varepsilon^2)b_n}^{\infty} u dF(u) \to 0.$$

Applying again Lemma 2.6 with $\delta = \varepsilon$, $x = (1-\varepsilon)\gamma_n(d_n)$ and $k = a_n$, we conclude that inequality (1.23) holds for all sufficiently large n.

Suppose that condition 2) is satisfied. Relations (3.8) and (3.9) imply that the conditions of Lemma 2.5 are fulfilled. By Lemma 2.5 with $y = y_n$, $k = a_n$, $h = h^*$ and $x = m_n(h^*)$, we have

$$\begin{aligned} P(S_{a_n} \ge (1-\varepsilon)^3 b_n) &\ge P(S_{a_n} \ge (1-\varepsilon)^2 a_n \gamma_n((1-\varepsilon)d_n)) \\ &= P(S_{a_n} \ge (1-\varepsilon)^2 a_n m_n(h^*)) \ge \frac{3}{4} e^{-(1-\varepsilon)(1+\varepsilon)\beta_n} = \frac{3}{4} e^{-(1-\varepsilon^2)\beta_n} \end{aligned}$$

for all sufficiently large n and (1.23) holds.

Theorem 3.6 follows now from Theorem 1.2. □

Now we turn to lower bounds for sequences $\{a_n\}$ which increase slowly enough.

Theorem 3.7. *If the conditions of Theorem 3.6 hold and* $\log\log n = o(\log(n/a_n))$, *then*

$$\liminf \frac{U_n}{b_n} \ge 1 \quad a.s.$$

We checked above that under the assumptions of Theorem 3.6, inequality (1.23) holds. Hence, Theorem 3.7 follows from Theorem 1.3.

Theorems 3.5–3.7 and the inequalities $R_n \le U_n \le W_n$ yield the next result.

Theorem 3.8. *If the conditions of Theorems 3.5 and 3.6 hold, then*

$$\limsup \frac{W_n}{b_n} = \limsup \frac{U_n}{b_n} = \limsup \frac{R_n}{b_n} = \limsup \frac{T_n}{b_n} = 1 \quad a.s.$$

If $\log\log n = o(\log(n/a_n))$ *in addition, then*

$$\lim \frac{W_n}{b_n} = \lim \frac{U_n}{b_n} = 1 \quad a.s.$$

3.4 Corollaries of the Universal Strong Laws

We derive various corollaries of Theorems 3.4 and 3.8 in this section. The results crucially depend on the growth rate of a_n. We start with short increments when $a_n = O(\log n)$. In this case, the norming sequences depend on the full distribution of X and, sometimes, determine this distribution uniquely. Such results are called the Erdős–Rényi and Shepp laws. Further, we turn to large increments when $a_n/\log n \to \infty$. In this case, the norming sequences depend on parameters of the completely asymmetric stable laws which domains of attraction distributions of summands belong to. For example, $EX = 0$ and $EX^2 = 1$ (i.e. $F \in DN(2)$) are necessary and sufficient conditions for the LIL with the norming sequence $\sqrt{2n \log\log n}$. Analogues of the LIL for $F \in DN(\alpha)$, $\alpha \in (1, 2)$, hold with the norming sequences $C(\alpha)n^{1/\alpha}(\log\log n)^{1-1/\alpha}$. The non-normal attraction requires an additional slowly varying multiplier in the normalization. We show that the situation is similar for large increments. Results for them are called the Csörgő–Révész laws. They also imply some special cases of the LIL that are discussed as well. Finally, we discuss the SLLN.

3.4.1 *The Erdős–Rényi and Shepp Laws*

Consider short increments including the case $a_n = 1$ for all n. Start with $a_n = c \log n$.

Theorem 3.9. *If $EX \geq 0$, $h_0 > 0$ and $a_n = [c \log n]$, $c > 0$, then*

$$\lim \frac{W_n}{a_n} = \lim \frac{U_n}{a_n} = \limsup \frac{T_n}{a_n} = \limsup \frac{R_n}{a_n} = \gamma\left(\frac{1}{c}\right) \quad a.s.$$

Theorem 3.9 follows from Theorem 3.4. It is the full form of the Erdős–Rényi law (for U_n and W_n) and the Shepp law (for T_n).

The Erdős–Rényi and Shepp laws are quite different from the SLLN and LIL. For example, the norming sequences depend in the last strong laws on first or second moments of X for $F \in DN(2)$. To find $\gamma(x)$, we have to know the full distribution of X. Moreover, $\gamma(x)$ defines uniquely this distribution provided the two-sided Cramér condition holds (see Section 2.5 for details). In particular, if X is a standard normal random variable, then $\gamma(x) = \sqrt{2x}$ and the limit in Theorem 3.9 is $\sqrt{2/c}$ for all $c > 0$. The latter may hold for the normal case only.

The Erdős–Rényi and Shepp laws yield that the best accuracy of approximation of sums S_n by the Wiener process in the strong invariance principle is $O(\log n)$.

We now turn to the case $a_n = o(\log n)$.

Theorem 3.10. *Assume that $EX \geq 0$, $h_0 > 0$ and $a_n = o(\log n)$. For $\omega = \infty$, assume in addition that either condition 4), or condition 5) of Theorem 3.2 holds. Then*

$$\lim \frac{W_n}{a_n \gamma(\log n/a_n)} = \lim \frac{U_n}{a_n \gamma(\log n/a_n)} = \limsup \frac{T_n}{a_n \gamma(\log n/a_n)} = 1 \quad a.s.$$

One can replace T_n by R_n in the last relation.

Theorem 3.10 follows from Theorem 3.4. For U_n, Theorem 3.10 is Mason's extension of the Erdős–Rényi law.

If $a_n = 1$ for all n, then Theorem 3.4 yields

$$\lim \frac{1}{\gamma(\log n)} \max_{1 \leq k \leq n} X_k = 1 \quad \text{a.s.}$$

Hence, the norming is $\sqrt{2 \log n}$ for the standard normal random variables. If X has the exponential distribution with the density $p(x) = \mathrm{e}^{-x}$ for $x > 0$, then $\zeta(x) = x - 1 - \log x$ for $x \geq 1$. This implies that $\gamma(x) \sim x$ as $x \to \infty$. It follows that the norming for maxima is $\log n$ in the exponential case. We mentioned these results of the theory of extremes in Section 1.1.

3.4.2 *The Csörgő–Révész Laws*

The case $a_n/\log n \to \infty$ is quite different. The results below do not follows immediately from Theorems 3.4 and 3.8 as above. Their proofs additionally require complicated technical calculations.

We now present simple formulae for b_n under three one-sided moment conditions: the Cramér condition, the Linnik condition and $E(X^+)^p < \infty$. These conditions determine different minimal rates of the growth of a_n under consideration.

First, we deal with the case of finite variations.

Theorem 3.11. *Assume that $EX = 0$, $EX^2 = 1$ and one of the following conditions hold:*

1) $h_0 > 0$ and $a_n/\log n \to \infty$;

2) $\mathrm{E}e^{(X^+)^\beta} < \infty$ for some $\beta \in (0,1)$ and $a_n \geq C(\log n)^{2/\beta-1}$, where $C = C(\beta)$ is an absolute positive constant;

3) $E(X^+)^p < \infty$ for some $p > 2$ and $a_n \geq Cn^{2/p}/\log n$, where C is an arbitrary positive constant.

Then the relation

$$\limsup \frac{W_n}{b_n} = \limsup \frac{U_n}{b_n} = \limsup \frac{R_n}{b_n} = \limsup \frac{T_n}{b_n} = 1 \quad a.s. \tag{3.10}$$

holds with

$$b_n = \sqrt{2a_n \left(\log \frac{n}{a_n} + \log\log n \right)}.$$

If $\log\log n = o(\log(n/a_n))$ in addition, then

$$\lim \frac{W_n}{b_n} = \lim \frac{U_n}{b_n} = 1 \quad a.s. \tag{3.11}$$

We postpone the proof of Theorem 3.11 and join it with that in the case $DN(\alpha)$.

Theorem 3.11 generalizes the strong approximation laws obtained by M.Csörgő and P.Révész (see [Csörgő and Révész (1981)], Theorem 3.2.1). They have used the strong approximation of sums of independent random variables by the Wiener process. It is only possible under two-sided moment restrictions. Our assumptions are one-sided and we use an analysis of probabilities of large deviations in our proofs. This yields us to stronger results those are called the Csörgő–Révész laws.

If X has the standard normal distribution, then $\gamma(x) = \sqrt{2x}$ for all $x \geq 0$. Hence, the norming in Theorem 3.11 is the same as that in the

Gaussian case. In this sense, the result is invariant from the distribution of X. This distinguishes the Csörgő–Révész laws and the Erdős–Rényi law discussed above.

Turn to the case $EX^2 = \infty$.

Start with $F \in D(2)$. Note that $F \in DN(2)$ in Theorem 3.11.

Theorem 3.12. *Assume that $EX = 0$ and $F \in D(2)$. Put*

$$b_n = a_n \hat{m}(\hat{f}^{-1}(d_n)), \tag{3.12}$$

where

$$\hat{m}(h) = hG\left(\frac{1}{h}\right), \quad \hat{f}(h) = \frac{h^2}{2}G\left(\frac{1}{h}\right), \quad G(x) = \int_{-x}^{0} u^2 dF(u), \quad x > 0.$$

($G(x) \in SV_\infty$.)

Assume that one of the following conditions holds:

1) $h_0 > 0$ and $a_n/\log n \to \infty$;

2) $E\mathrm{e}^{(X^+)^\beta} < \infty$ for some $\beta \in (0,1)$ and $b_n \geq C(\log n)^{1/\beta}$, where $C = C(\beta)$ is an absolute positive constant;

3) $E(X^+)^p < \infty$ for some $p > 2$ and $b_n \geq Cn^{1/p}$, where C is an arbitrary positive constant.

Then relation (3.10) holds with b_n from (3.12). If $\log\log n = o(\log(n/a_n))$ in addition, then relation (3.11) holds with b_n from (3.12).

Below we prove Theorem 3.12 together with its analogue for $D(\alpha)$, $\alpha \in (1,2)$. In there, we check that $\hat{f}^{-1}(x) = \sqrt{2x}L(1/x)$, where $L(x) \in SV_\infty$. Hence, (3.12) is equivalent to

$$b_n = \frac{\sqrt{2a_n\beta_n}}{L(1/d_n)}. \tag{3.13}$$

Note that we give a way of calculation of the function $L(x)$ from $G(x)$. Comparing the last formula and the formula for b_n from Theorem 3.11 yields that there is a slowly varying multiplier in (3.13) which depends on the truncated second moment.

The inequalities for a_n from conditions 2) and 3) of Theorem 3.11 imply the inequalities $b_n \geq C_1(\log n)^{1/\beta}$ and $b_n \geq C_2 n^{1/p}$, correspondingly. It means that conditions 2) and 3) of Theorems 3.11 and 3.12 are similar. Unfortunately, the slowly varying multiplier in (3.13) does not allow to transform inequalities for b_n in simple inequalities for a_n in Theorem 3.12. Since $L(x) \to 0$ as $x \to \infty$, the conclusion of Theorem 3.12 holds if conditions 2) and 3) are replaced by conditions 2) and 3) of Theorem 3.11, correpondingly.

Turn to the case $F \in DN(\alpha)$, $\alpha \in (1, 2)$.

Theorem 3.13. *Assume that* $EX = 0$, $F \in DN(\alpha)$, $\alpha \in (1, 2)$ *and one of the following conditions hold:*

1) $h_0 > 0$ *and* $a_n / \log n \to \infty$;

2) $Ee^{(X^+)^\beta} < \infty$ *for some* $\beta \in (0, 1)$ *and* $a_n \geq C(\log n)^{\alpha/\beta - \alpha + 1}$, *where* $C = C(\alpha, \beta)$ *is an absolute positive constant;*

3) $E(X^+)^p < \infty$ *for some* $p > \alpha$ *and* $a_n \geq C n^{\alpha/p} / (\log n)^{\alpha - 1}$, *where* C *is an arbitrary positive constant.*

Then relation (3.10) holds with

$$b_n = \lambda^{-\lambda} a_n^{1/\alpha} \left(\log \frac{n}{a_n} + \log \log n \right)^\lambda, \tag{3.14}$$

where $\lambda = (\alpha - 1)/\alpha$.

If $\log \log n = o(\log(n/a_n))$ *in addition, then relation (3.11) holds with* b_n *from (3.14).*

Note that $b_n = \sqrt{2a_n(\log(n/a_n) + \log \log n)}$ provided we formally put $\alpha = 2$ in relation (3.14).

We now prove Theorems 3.11 and 3.13 both. The cases $\alpha = 2$ and $\alpha \in (1, 2)$ correspond to Theorems 3.11 and 3.13.

Proof. Assume that condition 1) is satisfied. Put $y_n = \infty$ and $h_n = (d_n/\lambda)^{1/\alpha}$. Applying Lemmas 2.8, 2.16 and 2.19 and relation (2.4), we get

$$\gamma_n(d_n) \sim \gamma_n(f_n(h_n)) \sim m_n(h_n) \sim h_n^{\alpha - 1} = \lambda^{-\lambda} d_n^\lambda. \tag{3.15}$$

It follows that one can replace (3.2) by (3.14).

Check the conditions of Theorem 3.4.

To this end, we need the next result.

Lemma 3.3. *Let* $V(x)$ *be a d.f. such that* $V(x)$ *is continuous at zero and* $V(0) < 1$. *Assume that* $P(X \geq 0) > 0$ *and there exists a sequence of positive constants* $\{B_n\}$ *such that the distributions of* S_n/B_n *converge weakly to the distribution* V. *Then there exists* $q > 0$ *such that* $P(S_n \geq 0) \geq q$ *for all* $n \in \mathbf{N}$.

Proof. Zero is a continuity point of $V(x)$. By the definition of the weak convergence, we have

$$P(S_n \geq 0) = P\left(\frac{S_n}{B_n} \geq 0\right) \to 1 - V(0).$$

Hence, there exists N_0 such that

$$P(S_n \geq 0) \geq \frac{(1 - V(0))}{2}$$

for all $n \geq N_0$. For $n < N_0$, we get

$$P(S_n \geq 0) \geq (P(X \geq 0))^n \geq (P(X \geq 0))^{N_0}.$$

Put $q = \min\{(1 - V(0))/2, (P(X \geq 0))^{N_0}\}$. The assumptions yield $q > 0$ and the result follows. □

By Lemma 3.3, relation (1.14) holds. It is clear that condition 1) of Theorem 3.2 is satisfied. Hence, Theorem 3.4 yields the result.

Suppose that condition 2) is fulfilled. Put $y_n = \lambda^{-\lambda} a_n^{1/\alpha} \beta_n^{\lambda}$ and $h_n = d_n^{1/\alpha}$.

If $a_n \geq C(\log n)^{\alpha/\beta - \alpha + 1}$, then, taking into account that $\beta_n \leq 2\log n$, we have

$$h_n y_n^{1-\beta} = \lambda^{\beta\lambda - 1} \beta_n^{1-\beta\lambda} a_n^{-\beta/\alpha} \leq \lambda^{\beta\lambda - 1} 2^{1-\lambda\beta} C^{-\beta/\alpha}.$$

Choosing large enough $C = C(\alpha, \beta)$, we have $\limsup h_n y_n^{1-\beta} < 1$. Making use of Lemmas 2.10, 2.17 and 2.19 and relation (2.4), we get (3.15). It follows that one can define b_n by formula (3.14).

Check now the conditions of Theorem 3.8.

Condition (1.12) follows from Lemma 3.1 and (3.14).

Taking into account that $a_n \geq C(\log n)^{\alpha/\beta - \alpha + 1}$ and the function $x(\log(n/x) + \log\log n)$ is increasing in $x \leq n$, we have

$$\begin{aligned} y_n = b_n &\geq \lambda^{-\lambda} \left(C(\log n)^{\alpha/\beta - \alpha + 1} \left(\log \left(\frac{n}{C(\log n)^{\alpha/\beta - \alpha}} \right) \right)^{\alpha - 1} \right)^{1/\alpha} \\ &\geq (2\lambda)^{-\lambda} C^{1/\alpha} (\log n)^{1/\beta} \geq (\log n)^{1/\beta} \end{aligned}$$

for all sufficiently large n provided C has been chosen large enough. Then

$$P(X \geq b_n) \leq P(\mathrm{e}^{(X^+)^\beta} \geq n)$$

for all sufficiently large n. It follows that the series from condition 1) of Theorem 3.5 converges.

Relation (1.14) follows from Lemma 3.3.

Relation (3.7) hods in view of $a_n/\log n \to \infty$. By Lemmas 2.10 and 2.17, we get

$$(h^*)^2 \sigma_n^2(h^*) = O(f_n(h^*)) = O(d_n). \tag{3.16}$$

Then (3.8) is equivalent to $1 = o(\sqrt{a_n d_n})$ that holds by $\beta_n \to \infty$.

Furthermore,

$$-EY_n = -\int_{-\infty}^{y_n} x dF(x) - y_n P(X \geq y_n) \leq \int_{y_n}^{\infty} x dF(x) \leq \int_{y_n}^{\infty} e^{x^\beta/2} dF(x)$$

$$\leq e^{-y_n^\beta/2} \int_0^\infty e^{x^\beta} dF(x) \leq \frac{Ee^{(X^+)^\beta}}{y_n}$$

for all sufficiently large n. Since $y_n = b_n$ and $b_n^{-1} = o(\gamma_n(d_n))$, relation (3.9) holds.

Then the conditions of Theorem 3.8 hold and we get the result.

Assume finally that condition 3) is satisfied. Put $y_n = (\lambda/d_n)^{1/\alpha}$ and $h_n = (d_n/\lambda)^{1/\alpha}$. By Lemmas 2.10, 2.17 and 2.19 and relation (2.4), we get (3.15). It follows that one can choose b_n from (3.14).

Check the conditions of Theorem 3.8. Condition (1.12) holds by Lemma 3.1 and (3.14).

Taking into account that $a_n \geq Cn^{\alpha/p}/(\log n)^{\alpha-1}$ and $x(\log(n/x) + \log\log n)$ is increasing in $x \leq n$, we have

$$b_n \geq \lambda^{-\lambda} \left(\frac{Cn^{\alpha/p}}{(\log n)^{\alpha-1}} \left(\log \frac{n(\log n)^\alpha}{Cn^{\alpha/p}} \right)^{\alpha-1} \right)^{1/\alpha} \geq C_1 n^{1/p},$$

for all sufficiently large n. This yields that

$$P(X \geq cb_n) \leq P\left(\left(\frac{X}{cC_1} \right)^p \geq n \right)$$

for all sufficiently large n. Hence, the series from condition 2) of Theorem 3.5 converges for all $c > 0$.

Applying the inequalities $a_n \geq Cn^{\alpha/p}/(\log n)^{\alpha-1}$ and $\beta_n \leq 2\log n$, we get

$$a_n P(X \geq y_n) \leq a_n \frac{E(X^+)^p}{y_n^p} \leq C_2 a_n^{1-p/\alpha} \beta_n^{p/\alpha} \leq C_3 n^{\alpha/p-1} (\log n)^{1-\alpha+p}$$

for all sufficiently large n. Further,

$$e^{-(1+\varepsilon)\beta_n} = a_n^{1+\varepsilon} (n \log n)^{-(1+\varepsilon)} \geq C_6 n^{(1+\varepsilon)(\alpha/p-1)} (\log n)^{-(1+\varepsilon)\alpha}.$$

It follows that

$$(a_n P(X \geq y_n))^2 = o\left(e^{-(1+\varepsilon)\beta_n} \right). \tag{3.17}$$

Note that the sequence $\{\delta h_n\}$, $\delta \in (0,1)$, satisfies to the conditions of Lemmas 2.10 and 2.17 as well as $\{h_n\}$. By Lemmas 2.10, 2.17 and 2.19 and relations (3.15) and (2.3), we have

$$\zeta_n(\delta\gamma_n(d_n)) \sim \zeta_n(\delta h_n^{\alpha-1}) \sim \zeta_n(m_n(\delta^{1/(\alpha-1)} h_n)) \sim f_n(\delta^{1/(\alpha-1)} h_n) \sim \delta^{1/\lambda} d_n.$$

Hence, for every $\mu \in (0,1)$, we write

$$\mathrm{e}^{(1+\varepsilon)\beta_n}\mathrm{e}^{-((1+\varepsilon)a_n-1)\zeta_n((1-\delta)\gamma_n(d_n))} \le \mathrm{e}^{\nu\beta_n} \le C_8 n^{-\nu(\alpha/p-1)}(\log n)^{C_9}$$

for all sufficiently large n, where $\nu = (1+\varepsilon)(1-(1-\delta)^{1/\lambda}(1-\mu))$. Take small enough δ and μ such that $\nu < 1$. Then $\nu(\alpha/p-1) > \alpha/p-1$. It follows that

$$a_n P(X \ge y_n)\mathrm{e}^{-((1+\varepsilon)a_n-1)\zeta_n((1-\delta)\gamma_n(d_n))} = o\left(\mathrm{e}^{-(1+\varepsilon)\beta_n}\right). \tag{3.18}$$

Taking into account (3.17), we get (3.6).

Relation (1.14) follows from Lemma 3.3.

Relation (3.7) holds for all sufficiently large n in view of $a_n/\log n \to \infty$. By Lemmas 2.10 and 2.17, we obtain (3.16). Then (3.8) holds again.

Since $E(X^+)^2 < \infty$, we have

$$-EY_n \le \int\limits_{y_n}^{\infty} x dF(x) = o\left(y_n^{-1}\right) = o\left(\gamma_n(d_n)\right).$$

This yields (3.9).

The result follows from Theorem 3.8. □

We now turn to domains of non-normal attraction.

Theorem 3.14. *Assume that $EX = 0$ and $F \in D(\alpha)$, $\alpha \in (1,2)$. Put*

$$b_n = a_n \hat{m}(\hat{f}^{-1}(d_n)), \tag{3.19}$$

where

$$\hat{m}(h) = \frac{\alpha\Gamma(2-\alpha)}{\alpha-1}h^{\alpha-1}G\left(\frac{1}{h}\right), \quad \hat{f}(h) = \Gamma(2-\alpha)h^{\alpha}G\left(\frac{1}{h}\right),$$

$G(x) = x^{\alpha}F(-x)$, $x > 0$. *($G(x) \in SV_{\infty}$.)*

Assume that one of the following conditions holds:

1) $h_0 > 0$ *and* $a_n/\log n \to \infty$;

2) $E\mathrm{e}^{(X^+)^{\beta}} < \infty$ *for some* $\beta \in (0,1)$ *and* $b_n \ge C(\log n)^{1/\beta}$, *where* $C = C(\beta)$ *is an absolute positive constant;*

3) $E(X^+)^p < \infty$ *for some* $p > \alpha$ *and* $b_n \ge Cn^{1/p}$, *where C is an arbitrary positive constant.*

Then relation (3.10) holds with b_n from (3.19) . If $\log\log n = o(\log(n/a_n))$ *in addition, then relation (3.11) holds with b_n from (3.19).*

We find in the proof of Theorem 3.14 that $\hat{f}^{-1}(x) = \lambda^{\lambda-1}x^{1/\alpha}L(1/x)$, where $L(x) \in SV_\infty$. Hence, relation (3.19) may be written as

$$b_n = \frac{\lambda^{-\lambda}a_n^{1/\alpha}\beta_n^\lambda}{L(1/d_n)}. \tag{3.20}$$

Note that we give a way for calculation of $L(x)$. Hence, for $F \in D(\alpha)$, the norming is distinguished from that for $F \in DN(\alpha)$ (see Theorem 3.13) by a multiplier which depends of the slowly varying part of the tail of the distribution F. Unlike the case $F \in D(2)$, it turns out that $L(x)$ may tend to either infinity, or zero as $x \to \infty$. It depends on the behaviour of $G(x)$. If $G(x) \to 0$ as $x \to \infty$, then $L(x) \to \infty$ as $x \to \infty$. If $G(x) \to \infty$ as $x \to \infty$, then $L(x) \to 0$ as $x \to \infty$. In the last case, conditions 2) and 3) of Theorem 3.13 imply conditions 2) and 3) of Theorem 3.14, correspondingly.

We now prove Theorems 3.12 and 3.14. The cases $\alpha = 2$ and $\alpha \in (1,2)$ correspond to Theorems 3.12 and 3.14.

Proof. For $x > 0$, put $G(x) = x^\alpha F(-x)$ for $\alpha \in (1,2)$ and $G(x) = \int_{-x}^{0} u^2 dF(u)$ for $\alpha = 2$. For $h > 0$, denote

$$\hat{m}(h) = c_1(\alpha)h^{\alpha-1}G\left(\frac{1}{h}\right), \quad \hat{f}(h) = c_2(\alpha)h^\alpha G\left(\frac{1}{h}\right),$$

where $c_1(2) = 1$, $c_2(2) = 1/2$, $c_1(\alpha) = c_2$ and $c_2(\alpha) = c_4$ with the constants c_2 and c_4 from Lemma 2.16.

Assume first that condition 1) holds. Put $y_n = \infty$, $h_n = \hat{f}^{-1}(d_n)$. Applying Lemmas 2.13, 2.16 and 2.19 and relation (2.4), we get

$$\begin{aligned}\gamma_n(d_n) &\sim \gamma_n(\hat{f}(h_n)) \sim \gamma_n(f_n(h_n)) \sim m_n(h_n)\\ &\sim \hat{m}(h_n) \sim \hat{m}(\hat{f}^{-1}(d_n)).\end{aligned} \tag{3.21}$$

By Lemma 2.20, we have

$$\hat{f}^{-1}(x) = c_3(\alpha)x^{1/\alpha}L\left(\frac{1}{x}\right), \tag{3.22}$$

where $c_3(\alpha) = \lambda^{-1/\alpha}$ and $L(x) \in SV_\infty$.

Relation (3.21) and the definition of $\hat{m}(h)$ yield that

$$\gamma_n(d_n) \sim c_1(\alpha)(c_3(\alpha))^{\alpha-1}d_n^\lambda\left(L\left(\frac{1}{d_n}\right)\right)^{\alpha-1}G\left(\frac{1}{c_3(\alpha)d_n^{1/\alpha}L(1/d_n)}\right).$$

Since

$$x \sim \hat{f}(\hat{f}^{-1}(x)) = c_4(\alpha)x\left(L\left(\frac{1}{x}\right)\right)^\alpha G\left(\frac{1}{c_3(\alpha)x^{1/\alpha}L(1/x)}\right)$$

as $x \to 0$, where $c_4(2) = 1$ and $c_4(\alpha) = c_4/\lambda$ for $\alpha \in (1, 2)$, we get

$$c_4(\alpha) G\left(\frac{1}{c_3(\alpha) x^{1/\alpha} L(1/x)}\right) \sim \left(L\left(\frac{1}{x}\right)\right)^{-\alpha}$$

as $x \to 0$. Note that $L(x) \to \infty$ as $x \to 0$ for $\alpha = 2$.

Hence, we have

$$\gamma_n(d_n) \sim \frac{\lambda^{-\lambda} d_n^{\lambda}}{L(1/d_n)}. \tag{3.23}$$

It follows that one can put

$$b_n = \frac{\lambda^{-\lambda} a_n^{1/\alpha} \beta_n^{\lambda}}{L(1/d_n)}. \tag{3.24}$$

Check the conditions of Theorem 3.4. Relation (1.14) follows from Lemma 3.3. It is clear that condition 1) of Theorem 3.2 is satisfied. By Theorem 3.4, we get the result under condition 1).

Suppose now that condition 2) holds. Put $y_n = a_n \hat{m}(\hat{f}^{-1}(d_n))$, $h_n = \hat{f}^{-1}(d_n)$.

Check that $\limsup h_n y_n^{1-\beta} < 1$. Note that $h\hat{m}(h) = \lambda^{-1}\hat{f}(h)$ and

$$h_n y_n = a_n h_n \hat{m}(h_n) = \lambda^{-1} a_n \hat{f}(h_n) \sim \lambda^{-1}\beta_n.$$

We have

$$h_n y_n^{1-\beta} \leq \frac{2\beta_n}{\lambda y_n^{\beta}} = \frac{2\beta_n}{\lambda b_n^{\beta}} \leq \frac{2\beta_n}{\lambda C^{\beta}\log n} \leq \frac{4}{\lambda C^{\beta}}$$

for all sufficiently large n. It follows that $\limsup h_n y_n^{1-\beta} < 1$ provided $C = C(\alpha, \beta)$ is chosen large enough.

By Lemmas 2.15, 2.17 and 2.19 and relation (2.4), we get (3.21) and (3.23). It follows that the norming b_n may be taken from (3.24).

Check the conditions of Theorem 3.8. Remember that one can replace condition 1) by condition (1.21) of Remark 1.1 in Theorem 1.1 and, consequently, in Theorem 3.5. Note that inequalities (1.14) for S_i in Remark 1.1 follow from Lemma 3.3.

Put $b_n' = \lambda^{-\lambda} a_n^{1/\alpha} \beta_n^{\lambda}$ and $n_k = [\theta^k]$ for $k \in \mathbf{N}$. We have

$$\frac{b_{n_{k+1}}}{b_{n_k}} \sim \frac{b'_{n_{k+1}}}{b'_{n_k}} \frac{L(1/d_{n_k})}{L(1/d_{n_{k+1}})} = \frac{b'_{n_{k+1}}}{b'_{n_k}} \frac{L(1/d_{n_k})}{L((1/d_{n_k})\theta(1+o(1)))} \sim \frac{b'_{n_{k+1}}}{b'_{n_k}} \tag{3.25}$$

as $k \to \infty$ by the uniform convergence theorem (Theorem 1.1, p. 2 from [Seneta (1976)]).

Assume that $a_n/n \to 1$. Then

$$b_n \sim \lambda^{-\lambda}(\log\log n)\left(\frac{n}{\log\log n}\right)^{1/\alpha} L\left(\frac{n}{\log\log n}\right).$$

The function $x^{1/\alpha}L(1/x)$ is a regularly varying function of positive order and, hence, it is equivalent to a non-decreasing regularly varying function (see [Seneta (1976)], p. 20). Since the sequence $\{n/\log\log n\}$ is non-decreasing, $\{b_n\}$ is equivalent to a non-decreasing sequence. Relation (1.12) follows from (3.25) and Lemma 3.1. Then condition 1) of Theorem 1.1 holds.

Suppose that a_n/n is separated from 1. Proving Theorems 3.11 and 3.13, we checked that $\{b'_n\}$ is equivalent to a non-decreasing sequence and (1.12) holds for $\{b'_n\}$. Define n_k by formula (1.17). Then $\theta^{k-1} < n_{k-1} \le \theta^k < n_k \le \theta^{k+1}$. It follows that (1.12) holds for $\{b'_n\}$ with n_k instead of $[\theta^k]$ as well. Hence,

$$1 \le \frac{b_{n_k}}{m_k} \le \theta \frac{b'_{n_k} L'(1/d_{n_k})}{b'_{n_{k-1}} \min\limits_{n_{k-1}\le n\le n_k} L'(1/d_n)} \le \theta^2 \frac{L'(1/d_{n_k})}{\min\limits_{n_{k-1}\le n\le n_k} L'(1/d_n)} \tag{3.26}$$

for all sufficiently large k, where $L'(x) = 1/L(x)$.

Assume that $n_{k-1} \le n \le n_k$. Taking into account that a_n and β_n are non-decreasing and the inequality $x/y \ge \log x/\log y$ holds for $x \ge y \ge e$, we have

$$\frac{a_{n_k}\beta_n}{a_n\beta_{n_k}} \ge \frac{\beta_{n_{k-1}}}{\beta_{n_k}} = \frac{\log(n_{k-1}\log n_{k-1}/a_{n_{k-1}})}{\log(n_k \log n_k/a_{n_k})} \ge \frac{n_{k-1}\log n_{k-1}}{n_k \log n_k} \ge \frac{1}{\theta^3}$$

for all sufficiently large k. In the last inequality, we have used again that $\theta^{k-1} < n_{k-1} \le \theta^k < n_k \le \theta^{k+1}$ by (1.17). From the other hand, since $x/a(x)$ is non-decreasing, we get

$$\frac{a_{n_k}\beta_n}{a_n\beta_{n_k}} \le \frac{a_{n_k}}{a_{n_{k-1}}} \le \frac{n_k}{n_{k-1}} \le \theta^3$$

for all sufficiently large k. It follows that

$$\min_{n_{k-1}\le n\le n_k} L'\left(\frac{1}{d_n}\right) = L'\left(\frac{\tilde{\theta}_k}{d_{n_k}}\right),$$

where $\theta^{-3} \le \tilde{\theta}_k \le \theta^3$ for all sufficiently large k. Using again the uniform convergence theorem, we conclude that

$$\frac{L'(1/d_{n_k})}{L'(\tilde{\theta}_k/d_{n_k})} \to 1$$

as $k \to \infty$. This and (3.26) imply (1.21).

Hence, for $a_n/n \to 1$, condition 1) of Theorem 1.1 holds and for a_n/n separated from 1, condition (1.21) from Remark 1.1 holds.

Relation $y_n = b_n \geq (\log n)^{1/\beta}$ implies that the series in condition 1) of Theorem 3.5 converges.

Relation (1.14) follows from Lemma 3.3.

From Lemmas 2.15 and 2.17, we obtain (3.16) that implies (3.8). In the same way as in the proof of Theorems 3.11 and 3.13 for the case 2), we have $-EY_n = o(\gamma_n(d_n))$. Then we get (3.9).

By Theorem 3.8, we obtain the result in the case 2).

Assume now that condition 3) holds. Put $y_n = 1/\hat{f}^{-1}(d_n)$, $h_n = \hat{f}^{-1}(d_n)$.

Applying Lemmas 2.15, 2.17 and 2.19 and relation (2.4), we prove (3.21) and (3.23). The last relation yields (3.24).

In the same way as for the case 2), we prove that if $a_n/n \to 1$, then condition 1) of Theorem 1.1 holds and if a_n/n is separated from 1, then condition (1.21) from Remark 1.1 holds.

The inequality $b_n \geq Cn^{1/p}$ implies the convergence of the series from condition 2) of Theorem 3.5.

In the same way as for the case 3) of Theorems 3.11 and 3.13, we get (3.17) and (3.18). The last two relations yield (3.6).

Relation (1.14) follows by Lemma 3.3. We check (3.8) in the same way as for the case 2).

We get relation (3.9) in the same way as in the proof of the case 3) of Theorems 3.11 and 3.13.

By Theorem 3.8, we have the result under condition 3).

Theorems 3.12 and 3.14 are proved. □

3.4.3 *The Law of the Iterated Logarithm*

Turn to the LIL for increments. We start with simple corollaries of Theorems 3.11–3.14. We consequently consider the cases of finite variations, a domain of attraction of the normal law and domains of normal and non-normal attraction of the asymmetric stable laws.

Corollary 3.1. *Assume that $EX = 0$, $EX^2 = 1$, $E(X^+)^p < \infty$ for some $p > 2$ and*

$$\frac{\log(n/a_n)}{\log\log n} \to q \geq 0. \tag{3.27}$$

Then

$$\limsup \frac{W_n}{\sqrt{2a_n \log\log n}} = \limsup \frac{U_n}{\sqrt{2a_n \log\log n}} = \sqrt{1+q} \quad a.s. \tag{3.28}$$

Corollary 3.1 follows from Theorem 3.11 and (3.27). For $a_n = n$, $q = 0$ and $EX^2 = 1$, Corollary 3.1 yields the LIL for S_n and $\max_{1 \le k \le n} S_k$ with the standard norming $b_n = \sqrt{2n \log\log n}$.

Corollary 3.2. *Assume that* $EX = 0$, $F \in D(2)$, $E(X^+)^p < \infty$ *for some* $p > 2$ *and relation (3.27) holds. Then*

$$\limsup \frac{L(a_n/\log\log n)\, W_n}{\sqrt{2a_n \log\log n}} = \limsup \frac{L(a_n/\log\log n)\, U_n}{\sqrt{2a_n \log\log n}} = \sqrt{1+q} \quad a.s., \tag{3.29}$$

where $L(x)$ *is a slowly varying function from (3.13).*

Corollary 3.2 follows from Theorem 3.12 and relation (3.13).

Corollary 3.3. *Assume that* $EX = 0$, $F \in DN(\alpha)$, $\alpha \in (1, 2)$, *and* $E(X^+)^p < \infty$ *for some* $p > \alpha$. *Assume that (3.27) holds. Then*

$$\limsup \frac{W_n}{\lambda^{-\lambda} a_n^{1/\alpha} (\log\log n)^\lambda} = \limsup \frac{U_n}{\lambda^{-\lambda} a_n^{1/\alpha} (\log\log n)^\lambda} = (1+q)^\lambda \quad a.s., \tag{3.30}$$

where $\lambda = (\alpha - 1)/\alpha$.

Corollary 3.3 follows from Theorem 3.13 and (3.27). The next result follows from Theorem 3.14.

Corollary 3.4. *Assume that* $EX = 0$, $F \in D(\alpha)$, $\alpha \in (1, 2)$. *Assume that condition 3) of Theorem 3.14 and relation (3.27) hold. Then*

$$\limsup \frac{L(a_n/\log\log n) W_n}{\lambda^{-\lambda} a_n^{1/\alpha} (\log\log n)^\lambda} = \limsup \frac{L(a_n/\log\log n) U_n}{\lambda^{-\lambda} a_n^{1/\alpha} (\log\log n)^\lambda} = (1+q)^\lambda \quad a.s., \tag{3.31}$$

where $\lambda = (\alpha - 1)/\alpha$ *and* $L(x)$ *is a slowly varying function from (3.20).*

We dealt before with the moment conditions $E(X^+)^p < \infty$ for $p > \alpha$. We now turn to the case $p = \alpha$.

Theorem 3.15. *Assume that* $\alpha \in (1, 2]$, $EX = 0$ *and* $E(X^+)^\alpha (\log(1 + X^+))^\tau < \infty$ *for some* $\tau > 1/2$. *Suppose that relation (3.27) holds for* *some $q \in [0, 2\tau - 1)$ and $a_n \geq Cn(\log n)^{-\tau}(\log\log n)^{-(\alpha-1)}$, where C is an arbitrary positive constant.*

If $EX^2 = 1$, then (3.28) holds. If $F \in D(2)$, then (3.29) holds. If $F \in DN(\alpha)$, $\alpha \in (1,2)$, then (3.30) holds. If $F \in D(\alpha)$, $\alpha \in (1,2)$, then (3.31) holds.

For $a_n = n$, $q = 0$ and $EX^2 = 1$, we have the norming sequence $b_n = \sqrt{2n \log\log n}$ from the LIL while the last theorem yields the LIL for sums under $E(X^+)^2(\log(1+X^+))^\tau < \infty$, $\tau > 1/2$. This is a more restrictive than the Hartman–Wintner theorem. The above technique may be modified to the case of independent non-identically distributed random variables. Then one can derive a result for increments that implies the Hartman–Wintner theorem. See [Frolov (2004a)] for details.

Proof. Assume first that $F \in DN(\alpha)$, $\alpha \in (1,2]$. Put $h_n = (d_n/\lambda)^{1/\alpha}$ and $y_n = (\lambda/d_n)^{1/\alpha}$. Then Lemmas 2.10, 2.17 and 2.19 and relation (2.4) give $\gamma_n(d_n) \sim \lambda^{-\lambda} d_n^\lambda$ Hence, one can put

$$b_n = \lambda^{-\lambda}(1+q)^\lambda a_n^{1/\alpha}(\log\log n)^\lambda. \tag{3.32}$$

Check the conditions of Theorem 3.8.

Condition (1.12) holds by (3.32) and Lemma 3.1.

Put $H(x) = x^\alpha(\log x)^\tau$. It is not difficult to check that $H^{-1}(x) \sim \alpha^{\tau/\alpha} x^{1/\alpha}(\log x)^{-\tau/\alpha}$ as $x \to \infty$. The inequality $a_n \geq Cn(\log n)^{-\tau}(\log\log n)^{-\alpha+1}$ yields that $b_n \geq C_1 H^{-1}(n)$ for all sufficiently large n. Hence,

$$\sum_n P(X \geq cb_n) \leq \sum_n P\left(H\left(\frac{X}{C_2}\right) \geq n\right) < \infty$$

for all $c > 0$.

Further, we have

$$a_n P(X \geq y_n) \leq C_2 a_n y_n^{-\alpha}(\log y_n)^{-\tau} \leq C_3(\log n)^{-\tau}\log\log n$$

for all sufficiently large n.

For every $\varrho > 0$, the inequality $\mathrm{e}^{-(1+\varepsilon)\beta_n} \geq (\log n)^{-(1+\varepsilon)(1+\varrho)(1+q)}$ holds for all sufficiently large n. It follows that for all sufficiently small $\varepsilon > 0$, we get the relation

$$(a_n P(X \geq y_n))^2 = o\left(\mathrm{e}^{-(1+\varepsilon)\beta_n}\right).$$

For all positive $\mu < 1$ and $\delta < 1$, put $\nu = (1-\mu)(1-\delta)^{\alpha/(\alpha-1)}$. Then $\mathrm{e}^{-(1+\varepsilon)\beta_n + ((1+\varepsilon)a_n - 1)\zeta_n((1-\delta)\gamma_n(d_n)} \geq \mathrm{e}^{-(1+\varepsilon)\beta_n + (1+\varepsilon)\nu\beta_n}$

$\geq (\log n)^{-(1+\varepsilon)(1-\nu)(1+\varrho)(1+q)} \geq C_3(\log n)^{-\tau}\log\log n \geq a_n P(X \geq y_n)$

for all sufficiently large n provided μ, δ and ϱ are small enough. It follows that (3.6) holds.

The remain conditions of Theorem 3.8 can be verify in the same way as in the proof of Theorems 3.11 and 3.13. We finally need to apply Theorem 3.8.

Assume now that $F \in D(\alpha)$, $\alpha \in (1, 2]$. Put $h_n = \hat{f}^{-1}(d_n)$ and $y_n = 1/\hat{f}^{-1}(d_n)$.

By Lemmas 2.15, 2.17 and 2.19 and relation (2.4), we get

$$\gamma_n(d_n) \sim \frac{\lambda^{-\lambda} d_n^{\lambda}}{L(1/d_n)},$$

where $L(x)$ is a slowly varying function. Hence, we put

$$b_n = \frac{\lambda^{-\lambda}(1+q)^{\lambda} a_n^{1/\alpha} (\log\log n)^{\lambda}}{L(1/d_n)}.$$

We check all conditions of Theorem 3.8 besides (3.6) in the same way as in the proofs of Theorems 3.12 and 3.14. We only mention that

$$a_n P(X \geq y_n) \leq C_4 (\log n)^{-\tau} \log\log n \left(L\left(\frac{1}{d_n}\right)\right)^{\alpha}$$
$$\leq C_4 (\log n)^{-\tau+\varrho} \log\log n$$

for all sufficiently large n, where ϱ may be chosen arbitrary small. It yields that we keep the relations, obtained above in the proof of (3.6) for the case $F \in DN(\alpha)$.

We finally apply Theorem 3.8 to finish the proof. □

3.4.4 *The Strong Law of Large Numbers*

Assume now that $EX > 0$ and $a_n/\log n \to \infty$. Note that the case $a_n = O(\log n)$ is included in Theorems 3.9 and 3.10.

If $h_0 > 0$, then the definition of b_n and the equality $\gamma(0) = EX$ imply that $b_n \sim a_n EX$. Hence, Theorem 3.4 yields the SLLN for increments. Theorem 3.8 implies SLLN as well. These results are unified in the next theorem.

Theorem 3.16. *Assume that $EX > 0$ and $X - EX$ satisfies to the conditions of one of Theorems 3.11–3.15. Then*

$$\limsup \frac{W_n}{a_n} = \limsup \frac{U_n}{a_n} = \limsup \frac{R_n}{a_n} = \limsup \frac{T_n}{a_n} = EX \quad a.s.$$

If $\log\log n = o(\log(n/a_n))$ or $a_n = n$, then

$$\lim \frac{W_n}{a_n} = \lim \frac{U_n}{a_n} = EX \quad a.s.$$

We assume in Theorem 3.16 that $EX > 0$, but one can easily check that the theorem holds for W_n in the case $EX = 0$ as well. For other statistics, these relations holds for $EX \leq 0$, too.

Note that $R_n = S_n$ for $a_n = n$ and one can replace lim sup by lim in this case. One can prove the result for non-negative random variables only when S_n is non-decreasing. Hence, we get SLLN for sums. Further, $W_n = \max_{1\leq k\leq n} S_k$ for $a_n = n$ and maxima of sums and sums have the same behaviour.

Of course, the moment conditions in Theorem 3.16 are not optimal. Remember that the proof of the Kolmogorov SLLN essentially uses the formula for the norming sequence. This is impossible in the proof of the universal theorems since the structure of the norming sequence is more complicated. Moreover, the objects under investigation are more complicated than sums of independent random variables. We emphasize that we do not prove the SLLN and LIL separately. We derive them from the universal strong laws. The price for that is such non-optimality of moment assumptions in the SLLN and LIL for sums. Nevertheless, we will check in the next section that our moment assumptions in the theorems for increments are optimal.

Proof. Assume that $EX^2 = 1$. Put $h_n = (2d_n)^{1/2}$ and $y_n = (1/2d_n)^{1/2}$.

Denote $Y'_n = \min\{X - EX, y'_n\}$, $Z'_n = Y'_n - EY'_n$, $y'_n = y_n - EX$. It is clear that $Y_n = Z'_n + EY_n$. By Lemma 2.10, we have

$$m_n(h_n) = EY_n + h_n(1+o(1)), \quad \sigma_n^2(h_n) = 1+o(1), \quad f_n(h_n) = \frac{h_n^2}{2}(1+o(1)).$$

Taking into account that $EY_n \to EX$, we get $m_n(h_n) \to EX$. By (2.4), we can put $b_n = a_n EX$.

We check the conditions of Theorem 3.8 in the same way as in the proof of Theorem 3.11.

Assume that $X > 0$ and $a_n = n$. By Theorem 3.8, we have

$$\limsup \frac{S_n}{n} = EX \quad \text{a.s.} \tag{3.33}$$

The random variable $-X + 2EX$ is bounded and, therefore, it satisfies to the Cramér condition. Applying of Theorem 3.8 to $-X + 2EX$ yields (3.33) with lim inf instead of lim sup. It yields the SLLN for $X > 0$. Using the representation $X = X^+ - X^-$, we obtain the SLLN for arbitrary X.

The cases $F \in DN(\alpha)$, $\alpha \in (1, 2)$ and $F \in D(\alpha)$, $\alpha \in (1, 2]$, can be considered in the same way. We omit details. □

3.4.5 *Results for Moduli of Increments of Sums of Independent Random Variables*

All previous results of this section can not hold for moduli of increments of sums of i.i.d. random variables. For moduli of increments, the random variables X and $-X$ have to satisfy the one-sided moment conditions used above. The latter is only possible when $EX^2 < \infty$. In this case, we are able to derive corollaries from the previous results for

$$U_n^\star = \max_{0\le k\le n-a_n} |S_{k+a_n} - S_k|,$$
$$W_n^\star = \max_{0\le k\le n-a_n} \max_{1\le j\le a_n} |S_{k+j} - S_k|,$$
$$R_n^\star = |S_n - S_{n-a_n}|, \quad T_n^\star = |S_{n+a_n} - S_n|.$$

It is clear from the equality $|X| = \max\{X, -X\}$ that for instance, the norming for $W_n^\star$ has to be a maximum from the normings for W_n and the analogue of W_n constructed from $-X$'s. So, for short increments, we only can write the Erdős–Rényi law for X with a symmetrical distribution. The result is the same as Theorem 3.9. In non-symmetrical case, we can not specify the limit. Hence, we below deal with large increments only.

Our first result is the next theorem.

Theorem 3.17. *Assume that $EX = 0$, $EX^2 = 1$ and one of the following conditions hold:*

1) $Ee^{h^\star|X|} < \infty$ for some $h^\star > 0$ and $a_n/\log n \to \infty$;

2) $Ee^{|X|^\beta} < \infty$ for some $\beta \in (0,1)$ and $a_n \ge C(\log n)^{2/\beta - 1}$, where $C = C(\beta)$ is an absolute positive constant;

3) $E|X|^p < \infty$ for some $p > 2$ and $a_n \ge Cn^{2/p}/\log n$, where C is an arbitrary positive constant.

Then the relation

$$\limsup \frac{W_n^\star}{b_n} = \limsup \frac{U_n^\star}{b_n} = \limsup \frac{R_n^\star}{b_n} = \limsup \frac{T_n^\star}{b_n} = 1 \quad a.s. \tag{3.34}$$

holds with

$$b_n = \sqrt{2a_n\left(\log \frac{n}{a_n} + \log\log n\right)}.$$

If $\log\log n = o(\log(n/a_n))$ in addition, then

$$\lim \frac{W_n^\star}{b_n} = \lim \frac{U_n^\star}{b_n} = 1 \quad a.s. \tag{3.35}$$

Theorem 3.17 follows from Theorem 3.11. Indeed, the inequality $R_n^\star \leq U_n^\star \leq W_n^\star$ and the inequality

$$W_n^\star = \max\left\{W_n, \max_{0\leq k\leq n-a_n} \max_{1\leq j\leq a_n} (-S_{k+j}+S_k)\right\}$$

yield the upper bounds in (3.34) and (3.35). The lower bounds follow from the inequalities $R_n^\star \leq U_n^\star \leq W_n^\star$, $R_n \leq R_n^\star$, $U_n \leq U_n^\star$ and $T_n \leq T_n^\star$. To this end, the one-sided conditions of Theorem 3.11 is replaced in Theorem 3.17 by the two-sided ones. Hence, the Cramér condition, the Linnik condition and the power moment condition appear in 1)–3) correspondingly.

The LIL for moduli of increments is as follows.

Corollary 3.5. *Assume that $EX = 0$, $EX^2 = 1$, $E|X|^p < \infty$ for some $p > 2$ and*

$$\frac{\log(n/a_n)}{\log\log n} \to q \geq 0. \tag{3.36}$$

Then

$$\limsup \frac{W_n^\star}{\sqrt{2a_n \log\log n}} = \limsup \frac{U_n^\star}{\sqrt{2a_n \log\log n}} = \sqrt{1+q} \quad a.s. \tag{3.37}$$

Corollary 3.5 follows from Theorem 3.17 and (3.36). For $a_n = n$, $q = 0$ and $EX^2 = 1$, Corollary 3.5 yields the LIL for $|S_n|$ and $\max\limits_{1\leq k\leq n} |S_k|$ with the standard norming $b_n = \sqrt{2n\log\log n}$.

Reducing the range of $\{a_n\}$, we can relax the condition $E|X|^p < \infty$ for $p > 2$ in the LIL for moduli of increments of sums. In this case, we have the next better result.

Theorem 3.18. *Assume that $EX = 0$, $EX^2 = 1$ and $EX^2(\log(1+|X|))^\tau < \infty$ for some $\tau > 1/2$. Suppose that relation (3.36) holds for some $q \in [0, 2\tau - 1)$ and $a_n \geq Cn(\log n)^{-\tau}(\log\log n)^{-1}$, where C is an arbitrary positive constant.*

Then (3.37) holds.

Theorem 3.18 follows from Theorem 3.15 and the inequalities mentioned after Theorem 3.17.

We see that our techniques is sharp enough. We have the LIL under the "2 + log" moment condition instead of 2, but we keep a zone for the sequence $\{a_n\}$. Of course, the best condition is the existence of the second moment for $a_n = n$.

For SLLN, we have the following result.

Theorem 3.19. *Assume that either the conditions of Theorem 3.17, or the conditions of Theorem 3.18 hold. Then*

$$\lim \frac{W_n^\star}{a_n} = \lim \frac{U_n^\star}{a_n} = \lim \frac{R_n^\star}{a_n} = \lim \frac{T_n^\star}{a_n} = 0 \quad a.s.$$

Since considered functionals are non-negative, Theorem 3.19 follows from Theorems 3.17 and 3.18 and the relation $a_n/\log n \to \infty$ which implies that $b_n = o(a_n)$.

3.5 Optimality of Moment Assumptions

We first prove the necessity of the one-sided moment conditions for the Csörgő–Révész laws. We start with the Cramér condition.

Theorem 3.20. *Assume that*

$$\limsup \frac{W_n}{b_n} = 1 \quad a.s. \tag{3.38}$$

for all sequences $\{a_n\}$ *with* $a_n/\log n \to \infty$, *where*

$$b_n = \lambda^{-\lambda} a_n^{1/\alpha} \beta_n^\lambda L\left(\frac{a_n}{\beta_n}\right), \tag{3.39}$$

$L(x) \in SV_\infty$, $\alpha \in (1, 2]$, $\lambda = (\alpha - 1)/\alpha$. *Then* $h_0 > 0$.

Proof. We need the next result.

Lemma 3.4. *Let* X *be a random variable with* $P(X \geq 0) = 1$. *Assume that* $\mathrm{E}e^{X/g(X)} < \infty$ *for every non-decreasing function* $g(x)$ *with* $g(x) \to \infty$ *as* $x \to \infty$. *Then* $h_0 > 0$.

Proof. Assume that $Ee^{hx} = \infty$ for all $h > 0$.

Let $\{h_n\}$ be a strictly decreasing sequence of positive numbers such that $h_n \to 0$. Define a sequence of positive numbers $\{x_n\}$ as follows. Take $x_1 > 0$ such that $Ee^{h_1 X} I_{\{X<x_1\}} > 2$. If x_k is chosen, then we take x_{k+1} such that

$$Ee^{h_{k+1}X} I_{\{X<x_{k+1}\}} > 2Ee^{h_k X} I_{\{X<x_k\}}.$$

Since $h_{k+1} < h_k$, we have

$$Ee^{h_{k+1}X} I_{\{X<x_{k+1}\}} > 2Ee^{h_{k+1}X} I_{\{X<x_k\}}.$$

Hence, $x_{k+1} > x_k$.

Check that $x_k \to \infty$ as $x \to \infty$. Assume that $x_k \le C$. Then

$$2^k < E e^{h_k X} I_{\{X < x_k\}} \le E e^{h_k X} I_{\{X \le C\}}.$$

The last expectation is bounded for all sufficiently large k in view of $h_k \to 0$ as $k \to \infty$. This contradiction implies that $x_k \to \infty$ as $x \to \infty$.

Put $g(x) = 1/h_{k+1}$ for $x \in [x_k, x_{k+1})$. Then

$$E e^{X/g(X)} I_{\{x_1 \le X < x_{n+1}\}} = \sum_{k=1}^{n} E e^{h_{k+1} X} I_{\{x_k \le X < x_{k+1}\}}$$

$$= \sum_{k=1}^{n} \left(E e^{h_{k+1} X} I_{\{X < x_{k+1}\}} - E e^{h_{k+1} X} I_{\{X < x_k\}} \right)$$

$$> \frac{1}{2} \sum_{k=1}^{n} E e^{h_{k+1} X} I_{\{X < x_{k+1}\}} > \sum_{k=1}^{n} 2^k.$$

It follows that $E e^{X/g(X)} = \infty$. The latter contradicts to our assumption and the lemma is proved. □

Without loss of generality, we suppose that $a(x)/x \to 0$ as $x \to \infty$. Relation (3.38) implies that

$$\limsup \frac{X_{n-a_n}}{b_n} \le 1 \quad \text{a.s.}$$

There exists $N \in \mathbf{N}$ such that $\{n - a_n\}_{n=N}^{\infty}$ include all natural numbers from $N - a_N$. Of course, some repetitions of numbers are possible, but only two equal numbers may follow successively. Put

$$A_n = \{X_{n-a_n} \ge (1+\varepsilon) b_n\}$$

for all $n \ge N$. From the sequence $\{A_n\}$, we form subsequences $\{A_{n'}\}$ and $\{A_{n''}\}$ such that $\{A_n\} = \{A_{n'}\} \cup \{A_{n''}\}$ and each of them consists from events, generated by random variables with different indices. By independence of $\{A_{n'}\}$ and the Borel–Cantelli lemma, the series $\sum_{n'} P(A_{n'})$ converges. Similarly, the series $\sum_{n''} P(A_{n''})$ converges as well. Hence, for every $\varepsilon > 0$, we have

$$\sum_{n=1}^{\infty} P(X \ge (1+\varepsilon) b_n) < \infty. \tag{3.40}$$

Assume first that $b_n = \lambda^{-\lambda} a_n^{1/\alpha} \beta_n^{\lambda}$. Write $a_n = \varrho(n) \log n$, where $\varrho(n) \to \infty$. We may consider $\varrho(n)$ such that $\varrho(n)$ and $n/\varrho(n)$ are non-decreasing.

If $\log(n/\varrho(n)) = o(\log n)$, then for every $\varepsilon \in (0,1)$, the inequality $\varrho(n) > n^{1-\varepsilon}$ holds for all sufficiently large n. Hence, $b_n \geq \lambda^{-\lambda} n^{(1-\varepsilon)/\alpha} (\log n)^{1/\alpha} (\log\log n)^{\lambda}$ for all sufficiently large n.

If $\log(n/\varrho(n)) \sim \log n$, then $b_n \geq \lambda^{-\lambda} (\varrho(n))^{1/\alpha} \log n(1 + o(1))$.

Hence, $b_n / \log n \to \infty$ and relation (3.40) yields the convergence of

$$\sum_n P(X \geq \varrho_1(n) \log n)$$

for every non-decreasing function $\varrho_1(x)$ with $\varrho_1(n) \to \infty$.

The function $H(x) = \varrho_1(x) \log x$ strictly increases. We write the inverse function $H^{-1}(x) = e^{x/g(x)}$. Since $x = H^{-1}(H(x)) = \exp\{\varrho_1(x) \log x / g(H(x))\}$, we see that $g(x)$ is a non-decreasing function and $g(x) \to \infty$ as $x \to \infty$.

It follows that the series

$$\sum_n P\left(\mathrm{e}^{X/g(X)} \geq n\right)$$

converges for every non-decreasing function $g(x)$ with $g(x) \to \infty$ as $x \to \infty$. Then $Ee^{X^+/g(X^+)} < \infty$ for every such function $g(x)$. By Lemma 3.4, the Cramér condition holds.

Assume now that $b_n = \lambda^{-\lambda} a_n^{1/\alpha} \beta_n^{\lambda} L(a_n/\beta_n)$, where $L(x) \in SV_\infty$ and $L(x) \to 0$ or ∞ as $x \to \infty$. Writing again $a_n = \varrho(n) \log n$, $\varrho(n) \to \infty$. we prove the convergence of the series

$$\sum_n P(X \geq \varrho_1(n) \log n)$$

for all non-decreasing functions $\varrho_1(x)$ with $\varrho_1(n) \to \infty$. This yields the result in the same way as before. □

Turn to the one-sided Linnik condition.

Theorem 3.21. *Assume that relation (3.38) holds with b_n from (3.39) for all sequences $\{a_n\}$ such that $b_n \geq C(\log n)^{1/\beta}$, where $C = C(\alpha, \beta) < 1$ is an absolute positive constant. Then $\mathrm{E}\mathrm{e}^{(X^+)^\beta} < \infty$.*

Proof. Take the sequence $\{a_n\}$ such that $b_n = C(\log n)^{1/\beta}$. Then $(1+\varepsilon)b_n \leq (\log n)^{1/\beta}$ for small ε and relation (3.40) implies that $E\mathrm{e}^{(X^+)^\beta} < \infty$. □

Remember that conditions 2) of Theorems 3.11 and 3.13 are equivalent to the condition $b_n \geq C(\log n)^{1/\beta}$ of Theorem 3.21. The constants are different in there, of course.

Now, we deal with one-sided power moment conditions.

Theorem 3.22. *Assume that relation (3.38) holds with b_n from (3.39) for all sequences $\{a_n\}$ such that $b_n \geq Cn^{1/p}$, where $p > \alpha$, $C > 0$. Then $E(X^+)^p < \infty$.*

Proof. Take the sequence $\{a_n\}$ such that $b_n = n^{1/p}$. Then relation (3.40) yields that $E(X^+)^p < \infty$. □

Note that conditions 3) of Theorems 3.11 and 3.13 are equivalent to the condition $b_n \geq Cn^{1/p}$ of Theorems 3.22.

We have checked that our theorems for increments imply the LIL. It is well known that the conditions $EX = 0$ and $EX^2 = 1$ are necessary and sufficient for the Hartman–Wintner LIL. The situation is similar for the domain of attractions under consideration. We summarize this in the next result.

Theorem 3.23. *Assume that the conditions of one of Theorems 3.20–3.22 hold. Then $EX = 0$.*

If $\alpha = 2$ and $L(x) \equiv 1$, then $EX^2 = 1$. If $F(-x) \in RV_\infty$, $\alpha = 2$ and $L(x) \to \infty$ as $x \to \infty$, then $F \in D(2)$.

If $F(-x) \in RV_\infty$, $\alpha \in (1,2)$ and $L(x) \equiv 1$, then $F \in DN(\alpha)$. If $F(-x) \in RV_\infty$, $\alpha \in (1,2)$, $L(x) \to \infty$ or $L(x) \to 0$ as $x \to \infty$, then $F \in D(\alpha)$.

To prove Theorem 3.23, we need the following three results. The first one is the converse of the Hartman–Wintner LIL from [Martikainen (1980)] and [Rosalsky (1980)].

Theorem 3.24. *The relation*

$$\limsup \frac{S_n}{\sqrt{2n \log\log n}} = 1 \quad a.s.$$

holds, if and only if $EX = 0$ and $EX^2 = 1$.

The second result describes the a.s. behaviour of the random walk $\{S_n\}$ when EX does not exist. It has been obtained in [Kesten (1970)] .

Theorem 3.25. *If $EX^+ = EX^- = \infty$, then one of the following three relations holds:*

1) $\lim S_n/n = +\infty$ *a.s.*

2) $\lim S_n/n = -\infty$ *a.s.*

3) $\limsup S_n/n = +\infty$ *a.s.*, $\liminf S_n/n = -\infty$ *a.s.*

The third result has been proved in [Klass (1976)].

Theorem 3.26. *Assume that $EX = 0$. For every $y > 0$, put*

$$k(y) = \frac{y^2}{\int_0^y E|X|I_{\{|X|>u\}}du}.$$

Let $K(x)$ be the inverse function to $k(x)$. Put $c_n = K(n/\log\log n)\log\log n$ for $n \geqslant 3$.

Then

$$\limsup \frac{S_n}{c_n} \geqslant 1 \quad a.s.$$

Turn to the proof of Theorem 3.23.

Proof. Assume that $a_n = n$ for all n. We checked in the proofs of Theorem 3.11–3.14 that $\{b_n\}$ is equivalent to a non-decreasing sequence when $a_n/n \to 1$. Without loss of generality, we further assume that $\{b_n\}$ is non-decreasing.

It is clear that there exists a sequence of random indices $\{k_n\}$ such that $k_n \le n$ and

$$W_n = \max_{1\le k\le n} S_k = S_{k_n}.$$

Since b_n is non-decreasing, we get

$$\limsup \frac{W_n}{b_n} = \limsup \frac{S_{k_n}}{b_n} \le \limsup \frac{S_{k_n}}{b_{k_n}} \le \limsup \frac{S_n}{b_n} \quad \text{a.s.}$$

From the other hand, the inequality $W_n \ge S_n$ holds. It follows that

$$\limsup \frac{S_n}{b_n} = 1 \quad \text{a.s.} \tag{3.41}$$

If $\alpha = 2$ and $L(x) \equiv 1$, then $b_n = \sqrt{2n\log\log n}$. In this case, the relations $EX = 0$ and $EX^2 = 1$ follow from Theorem 3.24.

Relation (3.41) implies that

$$\limsup \frac{S_n}{n} = 0 \quad \text{a.s.} \tag{3.42}$$

Hence, Theorem 3.25 yields that $EX^+ < \infty$, or $EX^- < \infty$, or these two relations hold both. If $EX \neq 0$, then (3.42) contradicts to the SLLN. It follows that $EX = 0$.

By Theorem 3.26, we have

$$\limsup \frac{S_n}{c_n} \ge 1 \quad \text{a.s.}$$

This and relation (3.41) imply that

$$\liminf \frac{b_n}{c_n} \geq 1.$$

Hence, there exists a sequence $\{n_k\}$ such that $n_k \nearrow \infty$ as $k \to \infty$ and $c_{n_k} \leq 2b_{n_k}$. Note that

$$b_n = \lambda^{-\lambda} n^{1/\alpha} (\log\log n)^{\lambda} L\left(\frac{n}{\log\log n}\right).$$

It follows that for every $\varepsilon > 0$, the inequalities

$$K\left(\frac{n_k}{\log\log n_k}\right) \leq 2\lambda^{-\lambda}\left(\frac{n_k}{\log\log n_k}\right)^{1/\alpha} L\left(\frac{n_k}{\log\log n_k}\right)$$
$$\leq 2\lambda^{-\lambda}\left(\frac{n_k}{\log\log n_k}\right)^{1/(\alpha-\varepsilon)}$$

hold for all sufficiently large k. Put $y_n = K(n/\log\log n)$. Then

$$k(y_{n_k}) \geq \frac{y_{n_k}^{\alpha-\varepsilon}}{(2\lambda^{-\lambda})^{\alpha-\varepsilon}}$$

for all sufficiently large k. The definition of $k(y)$ and the last inequality yield that

$$\int_0^{y_{n_k}} E|X| I_{\{|X|>u\}} du \leq (2\lambda^{-\lambda})^{-(\alpha-\varepsilon)} y_{n_k}^{2-\alpha+\varepsilon}$$

for all sufficiently large k. Taking into account that

$$\int_0^y E|X| I_{\{|X|>u\}} du = \int_{-y}^y u^2 dF(u) + y^2 P(|X| > y) + y \int_y^\infty P(|X| > y) du,$$

we have

$$P(|X| > y_{n_k}) \leq (2\lambda^{-\lambda})^{-(\alpha-\varepsilon)} y_{n_k}^{-\alpha+\varepsilon}$$

for all sufficiently large k.

Applying one of conditions $h_0 > 0$, $Ee^{(X^+)^\beta} < \infty$ or $E(X^+)^p < \infty$, we obtain that $P(X > y) = o(P(X < -y))$ as $y \to \infty$. Hence, $P(|X| > y)$ is a regularly varying function. Let ϱ be the order of $P(|X| > y)$. Then the last inequality implies that $\varrho \leq -\alpha$. If $\varrho < -\alpha$, then (3.41) holds with another norming sequences $\{b_n\}$ by Theorems 3.11—3.14. It follows that $\varrho = -\alpha$. Hence, F belongs to a domain of attraction of the stable law with the c.f. (2.2).

Assume that $\alpha < 2$. Assume that $L(x) \equiv 1$. If F belongs to the domain of non-normal attraction, then (3.41) contradicts to Theorem 3.14. It follows that F belongs to the domain of normal attraction and $P(X < cx) \in DN(\alpha)$ for some $c > 0$. By Theorem 3.13, relation (3.41) holds with $b_n = c\lambda^{-\lambda} n^{1/\alpha} (\log\log n)^{\lambda}$. It yields that $c = 1$ and $F \in DN(\alpha)$. If $L(x) \to 0$ or $L(x) \to \infty$ as $x \to \infty$, then F can not belong to the domain of normal attraction by Theorem 3.13. It follows that $F \in D(\alpha)$.

Assume that $\alpha = 2$ and $L(x) \to \infty$ as $x \to \infty$. If F is from the domain of normal attraction of the normal law, then by Theorem 3.11, relation (3.41) holds with $b_n = c\sqrt{2n \log\log n}$ for some $c > 0$. This contradiction yields that $F \in D(2)$. □

We finish this section with the necessary condition for the Erdős–Rényi law. It turns out that it is the one-sided Cramér condition.

Theorem 3.27. *Assume that $a_n = [c \log n]$ for some $c > 0$ and (3.38) holds with $b_n = Ca_n$ for some $C > 0$. Then $h_0 > 0$.*

Proof. We get (3.40) in the same way as in the proof of Theorem 3.20. Since $b_n \sim cC \log n$, the latter yields $h_0 > 0$. □

3.6 Necessary and Sufficient Conditions for the Csörgő–Révész Laws

Results of the previous section allow us to write the results for large increments under necessary and sufficient conditions. We do that in this section. Statements of results are very simple for the case of finite variations.

We start with the strongest moment restrictions and the largest range of lengths of increments. In this case, Theorems 3.11, 3.13, 3.20 and 3.23 yield the following result.

Theorem 3.28. *Assume that $\alpha \in (1, 2]$.*

If $EX = 0$, $F \in DN(\alpha)$ and $h_0 > 0$, then

$$\limsup \frac{W_n}{\lambda^{-\lambda} a_n^{1/\alpha} \beta_n^{\lambda}} = 1 \quad a.s., \tag{3.43}$$

for every sequence $\{a_n\}$ with $a_n/\log n \to \infty$, where $\lambda = (\alpha - 1)/\alpha$. One can replace W_n by U_n, T_n and R_n in the last relation.

If $\log\log n = o(\log(n/a_n))$ in addition, then one can replace lim sup *by* lim *in (3.43). This remains true for U_n as well.*

Conversely, if $\alpha \in (1,2]$, relation (3.43) holds for every sequence $\{a_n\}$ with $a_n/\log n \to \infty$ and $F(-x) \in RV_\infty$, then $EX = 0$, $F \in DN(\alpha)$ and $h_0 > 0$. One can omit the regularity condition for $F(-x)$ in the case $\alpha = 2$.

For $\alpha = 2$, Theorem 3.28 yields the next result.

Corollary 3.6. *For every sequence $\{a_n\}$ with $a_n/\log n \to \infty$, the relation*

$$\limsup \frac{W_n}{\sqrt{2a_n\beta_n}} = 1 \quad a.s.,$$

holds if and only if $EX = 0$, $EX^2 = 1$ and $h_0 > 0$.

Now, we relax the moment assumptions and reduce the range of lengths of increments. Theorems 3.11, 3.13, 3.21 and 3.23 imply the next result.

Theorem 3.29. *Assume that $\alpha \in (1,2]$ and $\beta \in (0,1)$.*

If $EX = 0$, $F \in DN(\alpha)$ and $Ee^{(X^+)^\beta} < \infty$, then relation (3.43) holds for every sequence $\{a_n\}$ with $a_n \geq C(\log n)^{\alpha/\beta-\alpha+1}$, where $C = C(\alpha,\beta)$ is an absolute positive constant. One can replace W_n by U_n, T_n and R_n in (3.43).

If $\log\log n = o(\log(n/a_n))$ in addition, then one can replace lim sup *by* lim *in (3.43). This remains true for U_n as well.*

Conversely, if relation (3.43) holds for every sequence $\{a_n\}$ with $a_n \geq C(\log n)^{\alpha/\beta-\alpha+1}$ and $F(-x) \in RV_\infty$, then $EX = 0$, $F \in DN(\alpha)$ and $Ee^{t_0(X^+)^\beta} < \infty$ for some $t_0 > 0$. One can omit the regularity condition for $F(-x)$ in the case $\alpha = 2$.

Note that we do not specify the constant $C(\alpha,\beta)$ from the first part of Theorem 3.29. This is possible, but we only mention for simplicity that this constant exists. Hence, t_0 appears in the Linnik type condition in the converse part of Theorem 3.29. Of course, if C coincides with $C(\alpha,\beta)$ in the converse part, then one can put $t_0 = 1$.

For $\alpha = 2$, Theorem 3.29 implies the next result.

Corollary 3.7. *Assume that $\beta \in (0,1)$. For every sequence $\{a_n\}$ with $a_n \geq C(\log n)^{2/\beta-1}$, the relation*

$$\limsup \frac{W_n}{\sqrt{2a_n\beta_n}} = 1 \quad a.s.,$$

holds if and only if $EX = 0$, $EX^2 = 1$ and $Ee^{t_0(X^+)^\beta} < \infty$ for some $t_0 > 0$.

Turn to the weakest moment assumptions and narrowest range of lengths of increments. In this case Theorems 3.11, 3.13, 3.22 and 3.23 yield the following result.

Theorem 3.30. *Assume that $\alpha \in (1, 2]$ and $p > \alpha$.*

If $EX = 0$, $F \in DN(\alpha)$ and $E(X^+)^p < \infty$, then relation (3.43) holds for every sequence $\{a_n\}$ with $a_n \geq Cn^{\alpha/p}/(\log n)^{\alpha-1}$, where C is an arbitrary positive constant. One can replace W_n by U_n, T_n and R_n in (3.43).

If $\log\log n = o(\log(n/a_n))$ in addition, then one can replace lim sup *by* lim *in (3.43). This remains true for U_n as well.*

Conversely, if relation (3.43) holds for every sequence $\{a_n\}$ with $a_n \geq Cn^{\alpha/p}/(\log n)^{\alpha-1}$, $C > 0$, and $F(-x) \in RV_\infty$, then $EX = 0$, $F \in DN(\alpha)$ and $E(X^+)^p < \infty$. One can omit the regularity condition for $F(-x)$ in the case $\alpha = 2$.

For $\alpha = 2$, Theorem 3.30 gives the following result.

Corollary 3.8. *Assume that $p > 2$. For every sequence $\{a_n\}$ with $a_n \geq Cn^{2/p}/\log n$, the relation*

$$\limsup \frac{W_n}{\sqrt{2a_n\beta_n}} = 1 \quad a.s.,$$

holds if and only if $EX = 0$, $EX^2 = 1$ and $E(X^+)^p < \infty$.

Turn to the case $F \in D(\alpha)$. It is more complicated since a slowly varying multiplier appears in the norming sequence.

For the strongest moment conditions and widest range of $\{a_n\}$, we have the next result which follows from Theorems 3.12, 3.14, 3.20 and 3.23.

Theorem 3.31. *Assume that $\alpha \in (1, 2]$.*

If $EX = 0$, $F \in D(\alpha)$ and $h_0 > 0$, then

$$\limsup \frac{W_n}{\lambda^{-\lambda} a_n^{1/\alpha} \beta_n^\lambda L(a_n/\beta_n)} = 1 \quad a.s., \tag{3.44}$$

for every sequence $\{a_n\}$ with $a_n/\log n \to \infty$, where $L(x) \in SV_\infty$ and $\lambda = (\alpha-1)/\alpha$. One can replace W_n by U_n, T_n and R_n in the last relation.

If $\log\log n = o(\log(n/a_n))$ in addition, then one can replace lim sup *by* lim *in (3.44). This remains true for U_n as well.*

Conversely, if for every sequence $\{a_n\}$ with $a_n/\log n \to \infty$, relation (3.44) holds with some function $L(x) \in SV_\infty$, $L(x) \to 0$ or $L(x) \to \infty$, and $F(-x) \in RV_\infty$, then $EX = 0$, $F \in D(\alpha)$ and $h_0 > 0$.

Remember that in the first part of Theorem 3.31, the function $L(x)$ appears in relation (3.44) from the left tails of distribution of X for $\alpha < 2$ and the truncated second moment of X for $\alpha = 2$. In particular, $L(x)$ can not tend to zero for $\alpha = 2$.

Relaxing moment assumptions and narrowing the range of $\{a_n\}$, we get the following result from Theorems 3.12, 3.14, 3.21 and 3.23.

Theorem 3.32. *Assume that* $\alpha \in (1, 2]$ *and* $\beta \in (0, 1)$.

If $EX = 0$, $F \in D(\alpha)$ *and* $Ee^{(X^+)^\beta} < \infty$, *then (3.44) holds for every sequence* $\{a_n\}$ *with* $b_n \geq C(\log n)^{1/\beta}$, *where* b_n *is the norming from (3.44),* $L(x) \in SV_\infty$, $\lambda = (\alpha - 1)/\alpha$ *and* $C = C(\alpha, \beta)$. *One can replace* W_n *by* U_n, T_n *and* R_n *in the last relation.*

If $\log \log n = o(\log(n/a_n))$ *in addition, then one can replace* lim sup *by* lim *in (3.44). This remains true for* U_n *as well.*

Conversely, if for every sequence $\{a_n\}$ *with* $b_n \geq C(\log n)^{1/\beta}$, *relation (3.44) holds with some function* $L(x) \in SV_\infty$, $L(x) \to 0$ *or* $L(x) \to \infty$, *and* $F(-x) \in RV_\infty$, *then* $EX = 0$, $F \in D(\alpha)$ *and* $Ee^{t_0(X^+)^\beta} < \infty$ *for some* $t_0 > 0$.

One can check that the condition $b_n \geq C(\log n)^{1/\beta}$ turns to the condition of Theorem 3.29 provided we put $L(x) \equiv 1$ in the formula for b_n. In the case of arbitrary $L(x)$, one can not write a simple conditions on a_n.

Note that if C from the converse part coincides with $C(\alpha, \beta)$ from the first part, then $t_0 = 1$.

For the weakest moment assumptions and narrowest range of $\{a_n\}$, the result is as follows.

Theorem 3.33. *Assume that* $\alpha \in (1, 2]$ *and* $p > \alpha$.

If $EX = 0$, $F \in D(\alpha)$ *and* $E(X^+)^p < \infty$, *then (3.44) holds for every sequence* $\{a_n\}$ *with* $b_n \geq Cn^{1/p}$, *where* b_n *is the norming from (3.44),* $L(x) \in SV_\infty$, $\lambda = (\alpha - 1)/\alpha$ *and* $C > 0$. *One can replace* W_n *by* U_n, T_n *and* R_n *in the last relation.*

If $\log \log n = o(\log(n/a_n))$ *in addition, then one can replace* lim sup *by* lim *in (3.44). This remains true for* U_n *as well.*

Conversely, if for every sequence $\{a_n\}$ *with* $b_n \geq Cn^{1/p}$, *relation (3.44) holds with some function* $L(x) \in SV_\infty$, $L(x) \to 0$ *or* $L(x) \to \infty$, *and* $F(-x) \in RV_\infty$, *then* $EX = 0$, $F \in D(\alpha)$ *and* $Ee^{t_0(X^+)^\beta} < \infty$ *for some* $t_0 > 0$.

Theorem 3.33 follows from Theorems 3.12, 3.14, 3.22 and 3.23.

Note that the condition $b_n \geq Cn^{1/p}$ turns to the condition of Theorem 3.30 when we put $L(x) \equiv 1$ in the formula for b_n. One can not write a simple conditions on a_n for arbitrary function $L(x)$.

3.7 Bibliographical Notes

The first result for increments of sums of i.i.d. random variables has been obtained in [Shepp (1964)]. Nevertheless, the real interest to this topic started from the paper [Erdős and Rényi (1970)]. In [Csörgő (1979)], the Erdős–Rényi law has been proved under one-sided exponential moments. Full forms of the Erdős–Rényi and Shepp laws have been derived in [Deheuvels and Devroye (1987)]. The necessity of the Cramér condition for the Erdős–Rényi and Shepp laws has been proved in [Steinebach (1978)] and [Lynch (1983)] correspondingly. Bounds for convergence rates in the Erdős–Rényi and Shepp laws have been obtained in [Csörgő and Steinebach (1981)], [Deheuvels and Devroye (1987)]. The case $a_n = o(\log n)$ has been investigated in [Mason (1989)] and [Bacro and Brito (1991)].

Results for large increments may be found in [Csörgő and Révész (1981)]. Certain results of this type have been derived by [Book (1975b)] and [Frolov (1990)].

Necessity of two-sided moment conditions in the Csörgő–Révész results for moduli of increments of sums has been established in [Shao (1989)].

In [Lanzinger (2000)] and [Lanzinger and Stadtmüller (2000)], the behaviour of increments for $a_n = [c(\log n)^{2/p-1}]$ has been investigated under a Linnik type condition. This is a border case between the Erdős–Rényi law and the Csörgő–Révész laws. Similar result for random variables from $DN(\alpha)$ may be found in [Terterov (2011)].

Results for maxima similar to $\max\limits_{0\leq j\leq n} \max\limits_{0\leq k\leq n-j} (S_{k+j} - S_k)/f(k,n)$ has been proved in [Shao (1995)], [Steinebach (1998)] and [Lanzinger and Stadtmüller (2000)].

The behaviour of increments has been studied for non-negative random variables and random variables from the Feller class in [Einmahl and Mason (1996)]. The results are universal since minimal moment conditions are assumed while we suggest universality in another sense. Remember that our universal theorems work for a wider range for a_n under "fixed" moment assumptions.

M.Csörgő and P.Révész used the Komlós–Major–Tushnády strong approximation of sums by the Wiener process (see [Komlós *et al.* (1975,

1976)]) and own results for increments of the Wiener process.

We mentioned that the analysis of probabilities of large deviations is the strongest method of proofs of limit theorems on a.s. convergence.

This method has been used for investigations of the a.s. behaviour of increments of sums of i.i.d. random variables in [Frolov (1990, 2000, 1998, 2002c, 2003b,d)]. In there, one can find one-sided generalisations of the Csörgő–Révész results for random variables from domains of attraction of completely asymmetric stable laws. So, the universal theory has been built. It includes the SLLN, the LIL, the Erdős–Rényi and Shepp laws, the Mason extensions of the last laws and the Csörgő–Révész laws. Moment assumptions are either optimal, or close to optimal. The converses of the Csörgő–Révész laws have been derived in [Frolov (2005)].

For independent, non-identically distributed random variables, the Erdős–Rényi and Shepp laws are obtained in [Steinebach (1981)] and [Frolov (1993b)], correspondingly. Bounds for rates of convergence in these laws are derived in [Lin (1990)], [Frolov (1993b,a)] and [Frolov *et al.* (1997)]. For the Csörgő–Révész laws, one can confer [Book (1975b,a)], [Hanson and Russo (1985)], [Lin (1990)], [Lin *et al.* (1991)], [Cai (1992)], [Frolov (1991, 1993b, 2002a, 2004b)]. The universal strong laws are obtained in [Frolov (2004b)].

We now mention related results on LIL. Note that in this case, the large deviations method is the best as well. For the i.i.d. random variables, see, for example, [Klass (1976, 1977)] and [Pruitt (1981)]. Converses of the Hartman–Wintner theorem were independently obtained by [Martikainen (1980)], [Rosalsky (1980)] and [Pruitt (1981)]. LIL for asymmetric stable i.i.d. random variables may be found in [Mijnheer (1974, 1972)]. LIL for random variables from domain of attractions one can find in [Lipschutz (1956a)], [Brieman (1968)] and [Kalinauskaĭte (1971)].

One-sided LIL for increments of independent, non-identically distributed random variables is proved by [Frolov (2004a)]. For $a_n = n$, results from [Frolov (2004a)] imply those in [Martikainen (1985)] which, in turn, are generalizations of the Kolmogorov LIL [Kolmogorov (1929)] and the Hartman–Wintner LIL [Hartman and Wintner (1941)] both.

In this book, we deal with the i.i.d. case only. The results for non-identically distributed random variables allow to include the Kolmogorov LIL and the Hartman–Wintner LIL in the universal theory. Remember that the Hartman–Wintner theorem may be proved with an application of the Kolmogorov LIL for truncated from above random variables and the SLLN for sums of remainder parts of summands. The same holds true for increments as well. So, we can first build the universal theory for sums of independent, non-identically distributed random variables. Then we can finish the universal theory for i.i.d. case. This is done in [Frolov (2004b)]. In particular, a result of [Strassen (1964)] is derived for increments of sums of i.i.d. random variables.

Detailed bibliography on LIL and SLLN may be found in monographs [Petrov (1975, 1987, 1995)].

Chapter 4

Strong Limit Theorems for Processes with Independent Increments

Abstract. Universal strong laws are derived for homogeneous processes with independent increments. It follows the SLLN, the LIL, the Erdős–Rényi–Shepp laws and the Csörgő–Révész laws for such processes. The Wiener process, the Poisson process, the compound Poisson processes and stable processes are examples.

4.1 The Universal Strong Laws for Processes with Independent Increments

Let $\xi(t)$, $t \geq 0$, be a stochastically continuous, homogeneous process with independent increments such that $\xi(0) = 0$ a.s. and $\mu = E\xi(1) \in [0, \infty)$. There exists a modification of $\xi(t)$ with trajectories from the space of cádlàg functions on $[0, \infty)$ which we only deal with. Hence, we assume that $\xi(t)$ is right continuous and has left limits.

Let a_T be a non-decreasing function such that T/a_T is non-decreasing and $0 < a_T \leq T$. Put

$$W_T = \sup_{0 \leq t \leq T - a_T} \sup_{0 \leq s \leq a_T} (\xi(t+s) - \xi(t)),$$

$$U_T = \sup_{0 \leq t \leq T - a_T} (\xi(t + a_T) - \xi(t)),$$

$$Q_T = \xi(T + a_T) - \xi(T), \quad R_T = \xi(T) - \xi(T - a_T).$$

Note that $W_T = \sup\limits_{0 \leq s \leq T} \xi(s)$ and $U_T = R_T = \xi(T)$ for $a_T = T$.

For every continuous function c_T, the sets $\{U_T < c_T\}$, $\{U_T < c_T \text{ i.o.}\}$ etc. are events since they are determined by the process at rational points only. The same holds true for W_T, R_T, Q_T, $\sup\limits_{t \leq T} \xi(t)$ and $\inf\limits_{t \leq T} \xi(t)$.

In this section, we describe the set of norming (non-random) functions

b_T for which either

$$\limsup \frac{U_T}{b_T} = 1 \quad \text{a.s.},$$

or the last relation holds with lim instead of lim sup when it is possible. We assume in what follows that lim sup, lim inf, lim, O, o, $\to$ are taken as $T \to \infty$ if not pointed otherwise.

We follow the same pattern as that for sums of i.i.d. random variables. The only difference is that T is a continuous parameter.

Put $X = \xi(1)$ and $h_0 = \sup\{h : \varphi(h) = E\mathrm{e}^{hX} < \infty\}$.

If $h_0 > 0$, then we define the functions of the LDT $m(h)$, $\sigma^2(h)$, $f(h)$, $\zeta(x)$ and $\gamma(x)$ in the same way as in Chapter 3. We also put

$$\gamma_T(x) = \gamma(x), \qquad \text{for all } T \geq 0. \tag{4.1}$$

For $h_0 = 0$, take a continuous function y_T with $y_T \to \infty$. For all $T \geq 0$, we put

$$Y_T = \min\{X, y_T\}, \quad Z_T = \begin{cases} Y_T - EX, & \text{for} \quad \mu = 0, \\ \quad Y_T, & \text{for} \quad \mu > 0. \end{cases}$$

Then we define the functions of the LDT $m_T(h)$, $\sigma_T^2(h)$, $f_T(h)$, $\zeta_T(x)$ and $\gamma_T(x)$ for the random variable Z_T in the same way as before.

Put

$$b_T = a_T \gamma_T(d_T), \quad \text{where} \quad d_T = \frac{\beta_T}{a_T}, \quad \beta_T = \log\frac{T}{a_T} + \log\log\max(T, 3).$$

This formula is the main result of this chapter in fact. We will see below that it is the formula of norming sequences in strong limit theorems for processes with independent increments.

The first result is as follows.

Theorem 4.1. *Assume that b_T is equivalent to a continuous, non-decreasing function and*

$$\limsup_{\theta \searrow 1} \limsup_{T \to \infty} \frac{b_{\theta T}}{b_T} = 1. \tag{4.2}$$

Assume that one of the following conditions holds:

1) $h_0 > 0$.

2) $\sum_n P(\xi(1) \geq b_n) < \infty$ and for every small enough $\varepsilon > 0$, there exist $\tau > 0$ and $H \geq 0$ such that

$$P(\xi((1+\varepsilon)a_T) \geq (1+\varepsilon)b_T) \leq \mathrm{e}^{-(1+\tau)\beta_T} + Ha_T P(\xi(1) \geq b_T) \tag{4.3}$$

for all sufficiently large T.

Suppose that for every $\varepsilon > 0$, there exists $q \in (0,1)$ such that

$$P(\xi(t) \geq -\varepsilon b_T) \geq q \tag{4.4}$$

for all $t \leq (1+\varepsilon)a_T$ and all sufficiently large T.

Then

$$\limsup \frac{W_T}{b_T} \leq 1 \quad a.s. \tag{4.5}$$

For $a_T \to \infty$, one can replace $(1+\varepsilon)a_T$ by $[(1+\varepsilon)a_T]$ in (4.3). Here, $[\cdot]$ is the integer part of the number in brackets.

Proof. We need the next result.

Lemma 4.1. *Let $r, c \geq 0$ and $q > 0$. If $P(\xi(t) \geq -c) \geq q$ for all $t \leq T$, then*

$$P\left(\sup_{0 \leq s \leq t \leq T} (\xi(t) - \xi(s)) \geq r\right) \leq q^{-2} P(\xi(T) \geq r - 2c).$$

Proof. Put $t_{kn} = k2^{-n}T$ for $1 \leq k \leq 2^n$ and every $n \in \mathbf{N}$. The result follows from Lemma 1.1 and

$$P\left(\sup_{0 \leq s \leq t \leq T} (\xi(t) - \xi(s)) \geq r\right) = \lim_{n \to \infty} P\left(\sup_{0 \leq t_{kn} \leq t_{jn} \leq T} (\xi(t_{jn}) - \xi(t_{kn})) \geq r\right).$$

□

Assume that condition 2) holds. Take $\varepsilon \in (0,1)$. Put

$$A_T = \{W_T \geq (1+3\varepsilon)b_T\}$$

and $u_T = \varepsilon a_T$. By Lemma 4.1, we have

$$\begin{aligned}
P_T &= P(A_T) \\
&\leq P\left(\bigcup_{j=1}^{[T/u_T]+1} \left\{\sup_{(j-1)u_T \leq t \leq j u_T} \sup_{0 \leq s \leq a_T} (\xi(t+s) - \xi(t)) \geq (1+3\varepsilon)b_T\right\}\right) \\
&\leq \frac{(1+\varepsilon)T}{u_T} P\left(\sup_{0 \leq t \leq u_T} \sup_{0 \leq s \leq a_T} (\xi(t+s) - \xi(t)) \geq (1+3\varepsilon)b_T\right) \\
&\leq \frac{2T}{q^2 u_T} P(\xi((1+\varepsilon)a_T) \geq (1+\varepsilon)b_T).
\end{aligned} \tag{4.6}$$

for all sufficiently large T. By inequality (4.3), we get

$$\begin{aligned}
P_T &\leq \frac{2T}{q^2 u_T}\left(\mathrm{e}^{-(1+\tau)\beta_T} + H a_T P(\xi(1) \geq b_T)\right) \\
&\leq C_1 (\log T)^{-(1+\tau)} + C_2 T P(\xi(1) \geq b_T)
\end{aligned}$$

for all sufficiently large T.

Take $\theta > 1$. Put

$$n_k = \min\left\{n : \theta^{k-1} < n \le \theta^k, nP(\xi(1) \ge b_n) = \min_{\theta^{k-1}<m\le\theta^k} mP(\xi(1) \ge b_m)\right\}$$

for $k \in \mathbf{N}$. Then the series $\sum_k P_{n_k}$ converges (see the proof of Theorem 1.1). The Borel–Cantelli lemma yields

$$\limsup_{k\to\infty} \frac{W_{n_k}}{b_{n_k}} \le 1 + 3\varepsilon \quad \text{a.s.}$$

This inequality, relation (4.2) and the monotonicity of W_T imply (4.5).

Suppose now that $h_0 > 0$. Put $\varphi_t(h) = Ee^{h\xi(t)}$. It is not difficult to check that $\varphi_t(h) = (\varphi(h))^t$ for all $t > 0$ and $h \in (0, h_0)$. Indeed, for every naturals k and n,

$$\varphi_{k/n}(h) = (Ee^{h\xi(1/n)})^k = (Ee^{h\xi(1)})^{k/n} = (\varphi(h))^{k/n}.$$

The first equality follows by the independence of increments and the homogeneity of $\xi(t)$. The second equality is a result of an application of the first one with $k = n$. Then for rational t, we have $\varphi_t(h) = (\varphi(h))^t$. We get the last relation for all positive t by a limit passing over rational $\{t_n\}$ with $t_n \to t$ as $n \to \infty$. It yields that the function of large deviations and its inverse function for the random variable $\xi(t)$ are $t\zeta(z/t)$ and $t\gamma(x/t)$ correspondingly.

Take $\varepsilon \in (0, 2)$. Put $u_T = \varepsilon a_T$. By (4.6) and Tchebyshev's inequality, we get

$$P_T \le \frac{2T}{q^2 u_T} e^{-(1+\varepsilon)a_T\zeta(b_T/a_T)} = \frac{2T}{q^2 u_T} e^{-(1+\varepsilon)\beta_T}$$

for all sufficiently large T. Put $n_k = [\theta^k]$, where $\theta > 1$. Then the series $\sum_k P_{n_k}$ converges. The remainder of the proof coincides with that for the case 2).

For $a_T \to \infty$, we take $u_T = \varepsilon a_T - \{(1+\varepsilon)a_T\}$, where $\{\cdot\}$ is a fractional part of the number in brackets. The condition $a_T \to \infty$ yields that $u_T > 0$ for all sufficiently large T. The remainder of the proof is the same as that for the case 2). □

Our next result is the following theorem.

Theorem 4.2. *Assume that for every $\varepsilon > 0$ there exists $\tau > 0$ such that*

$$P(\xi(a_T) \ge (1-\varepsilon)b_T) \ge e^{-(1-\tau)\beta_T} \tag{4.7}$$

for all sufficiently large T. For $a_T/T \to 1$, assume additionally that conditions (4.2) and (4.4) are satisfied. Then

$$\limsup \frac{R_T}{b_T} \geq 1 \quad a.s. \tag{4.8}$$

One can replace R_T by Q_T in the last relation.

Proof. Take $\varepsilon \in (0,1)$. By (4.7), we get

$$P_T = P(R_T \geq (1-\varepsilon)b_T) = P(\xi(a_T) \geq (1-\varepsilon)b_T) \geq \mathrm{e}^{-(1-\tau)\beta_T}$$

for all sufficiently large T.

Suppose that $a_T/T \to \rho \in [0,1)$. Put $n_1 = 1$ and

$$n_{k+1} = \min\{n : \ n > n_k, \ n - a_n \geq n_k\}$$

for $k \in \mathbf{N}$. We have in the same way as in the proof of Theorem 1.2 that the series $\sum_k P_{n_k}$ diverges. Since the events

$$\{R_{n_k} \geq (1-\varepsilon)b_{n_k}\}$$

are independent, the Borel–Cantelli lemma yields (4.8).

Assume now that $a_T/T \to 1$. Take $\theta > 1$ and put $n_k = [\theta^k]$,

$$A_k = \{\xi(n_k) - \xi(n_{k-1}) \geq (1-2\varepsilon)b_{n_k}\},$$
$$D_k = \{\xi(n_{k-1}) - \xi(n_k - a_{n_k}) \geq -\varepsilon b_{n_k}\}$$

for $k \in \mathbf{N}$. Note that $a_{n_k} \sim (n_k - n_{k-1})\theta/(\theta-1)$ as $k \to \infty$. This and (4.4) imply that $P(D_k) \geq q$ for all sufficiently large k.

The following pairs of events are independent: A_k and D_k, A_k and $D_k\overline{A_{k-1}D_{k-1}}$, A_k and $D_k\overline{A_{k-1}D_{k-1}A_{k-2}D_{k-2}}, \ldots$

Choose T_k such that $a_{T_k} = n_k - n_{k-1}$, $k \in \mathbf{N}$. Then $T_k \sim a_{T_k} \sim n_k(\theta-1)/\theta$ as $k \to \infty$. If θ is large enough, then $(\theta-1)/\theta$ is close to 1 and we conclude by (4.2) that for every $\rho > 0$, the inequality $b_{n_k} \leq (1+\rho)b_{T_k}$ holds for all sufficiently large k. Take ρ such that $(1-2\varepsilon)(1+\rho) = 1-\varepsilon$. Taking into account (4.7), we have

$$P(A_k) \geq P(\xi(n_k - n_{k-1}) \geq (1-2\varepsilon)b_{n_k})$$
$$\geq P(\xi(a_{T_k}) \geq (1-\varepsilon)b_{T_k}) \geq \mathrm{e}^{-(1-\tau)\beta_{T_k}}$$

for all sufficiently large k. This yields that the series $\sum_k P(A_k)$ diverges.

Lemma 1.3 implies that

$$P(A_k D_k \text{ i.o.}) \geq q > 0.$$

Hence,

$$P(R_{n_k} \geq (1-3\varepsilon)b_{n_k} \text{ i.o.}) > 0.$$

By "0 or 1" law, the last probability equals to 1 and Theorem 4.2 follows for R_T. For Q_T, the proof follows a similar way. □

The next result is useful to check conditions (4.4) and (4.7).

Remark 4.1. 1) If $a_T > 1$ and $b_T \to \infty$, then one can check inequality (4.4) for $t = [(1+\varepsilon)a_T]$ only.

2) If $b_T \to \infty$, then one can replace $\xi(a_T)$ by $\xi([a_T])$ in inequality (4.7).

Proof. Assume that for every $\varepsilon > 0$ there exists $q \in (0,1)$ such that inequality (4.4) holds with $t = [(1+\varepsilon)a_T]$ for all $T \geq T_0$.

By

$$m_T = \min_{0 \leq k \leq [(1+\varepsilon)a_{T_0}]} P(\xi(k) \geq -\varepsilon b_T) \to 1,$$

we get $m_T \geq q$ for all $T \geq T_1 \geq T_0$.

Suppose that $T \geq T_1$. If $[(1+\varepsilon)a_{T_0}] \leq k \leq [(1+\varepsilon)a_T]$, then $k = [(1+\varepsilon)a_s]$ for some $s \in [T_1, T]$. Using that b_T is non-decreasing, we have

$$P(\xi(k) \geq -\varepsilon b_T) \geq P(\xi(k) \geq -\varepsilon b_s) \geq q.$$

Thus, (4.4) holds for $t = k = 1, 2, \ldots, [(1+\varepsilon)a_T]$.

Assume now that $k \leq t \leq k+1 \leq [(1+\varepsilon)a_T]+1$. Taking into account the independence of increments, homogeneity of the process and $b_T \to \infty$, we get

$$\begin{aligned}
P(\xi(t) \geq -2\varepsilon b_T) &\geq P(\xi(k) \geq -\varepsilon b_T) P(\xi(t) - \xi(k) \geq -\varepsilon b_T) \\
&\geq q P(\xi(t-k) \geq -\varepsilon b_T) \geq q P\left(\inf_{0 \leq s \leq 1} \xi(s) \geq -\varepsilon b_T\right) \geq 0.5q
\end{aligned}$$

for all $T \geq T_2 \geq T_1$.

The first assertion follows. Turn to the next one.

By the independence of increments and the homogeneity of $\xi(t)$ and $b_T \to \infty$, we have

$$\begin{aligned}
&P(\xi(a_T) \geq (1-2\varepsilon)b_T) \geq P(\xi([a_T]) \geq (1-\varepsilon)b_T) P(\xi(a_T) - \xi([a_T]) \geq -\varepsilon b_T) \\
&\geq P(\xi([a_T]) \geq (1-\varepsilon)b_T) P\left(\inf_{0 \leq s \leq 1} \xi(s) \geq -\varepsilon b_T\right) \\
&\geq 0.5 P(\xi([a_T]) \geq (1-\varepsilon)b_T)
\end{aligned}$$

for all sufficiently large T. □

The next result is as follows.

Theorem 4.3. *Assume that b_T is equivalent to a non-decreasing function, $b_T \to \infty$ and condition (4.2) is satisfied. Suppose that the conditions of Theorem 4.2 hold and* $\log\log T = o(\log(T/a_T))$. *Then*

$$\liminf \frac{U_T}{b_T} \geq 1 \quad a.s.$$

Proof. Assume that $j \le T < j+1$, where $j \ge 2$ and $j/a_j > 1$. Put $n_j = [j/a_j] - 1$,

$$I_k = \inf_{a_j \le s \le a_{j+1}} (\xi(ka_j + s) - \xi(ka_j))$$

for $k \in K$, where $K = \{0, 2, \ldots, 2([n_j/2] - 1)\}$. It is clear that

$$U_T \ge \max_{k \in K} I_k = M_j.$$

By $a_j/j \ge a_{j+1}/(j+1)$, we have $0 \le a_{j+1} - a_j \le a_j/j \le 1$ and $ka_j + a_{j+1} \le (k+1+1/j)a_j < (k+2)a_j$. Hence,

$$\begin{aligned} p_j &= P(M_j \le (1-2\varepsilon)b_j) = (P(I_0 \le (1-2\varepsilon)b_j))^{[n_j/2]} \\ &\le \mathrm{e}^{-[n_j/2]P(I_0 \ge (1-2\varepsilon)b_j)}. \end{aligned}$$

We have used here the independence of increments and the homogeneity of $\xi(t)$. Using these properties again, we get

$$\begin{aligned} &P(I_0 \ge (1-2\varepsilon)b_j) \\ &\ge P(\xi(a_j) \ge (1-\varepsilon)b_j)\, P\left(\inf_{a_j \le s \le a_{j+1}} (\xi(s) - \xi(a_j)) \ge -\varepsilon b_j\right) \\ &= P(\xi(a_j) \ge (1-\varepsilon)b_j) P\left(\inf_{0 \le s \le a_{j+1} - a_j} \xi(s) \ge -\varepsilon b_j\right) \\ &\ge P(\xi(a_j) \ge (1-\varepsilon)b_j) P\left(\inf_{0 \le s \le 1} \xi(s) \ge -\varepsilon b_j\right) \\ &\ge 0.5 P(\xi(a_j) \ge (1-\varepsilon)b_j) \end{aligned}$$

for all sufficiently large j. Taking into account (4.7), we obtain

$$p_j \le \exp\left\{-0.5 \left[\frac{n_j}{2}\right] \mathrm{e}^{-(1-\tau)\beta_j}\right\}$$

for all sufficiently large j. By $\log\log T = o(\log(T/a_T))$, the series $\sum_j p_j$ converges. The Borel–Cantelli lemma implies that with probability 1, the inequality $M_j > (1-2\varepsilon)b_j$ holds for all sufficiently large j. Since $b_{j+1} \sim b_j$ as $j \to \infty$, we conclude that with probability 1, the inequality $U_T > (1 - 3\varepsilon)b_T$ holds for all sufficiently large T. Theorem 4.3 follows. □

Theorems 4.1–4.3 and the inequalities $R_T \le U_T \le W_T$ yield the next result.

Theorem 4.4. *If the conditions of Theorems 4.1 and 4.2 hold, then*

$$\limsup \frac{W_T}{b_T} = \limsup \frac{U_T}{b_T} = \limsup \frac{R_T}{b_T} = \limsup \frac{Q_T}{b_T} = 1 \quad a.s. \tag{4.9}$$

If the conditions of Theorems 4.1 and 4.3 hold, then

$$\lim \frac{W_T}{b_T} = \lim \frac{U_T}{b_T} = 1 \quad a.s. \tag{4.10}$$

4.2 Strong Laws for Increments of Wiener and Stable Processes without Positive Jumps

In this section, we state corollaries of Theorem 4.4 for important processes with independent increments: the Wiener process and the stable processes without positive jumps. In the last case, increments are assumed to have asymmetric stable distributions with $\alpha \in (1, 2)$. For these processes, norming functions can be calculated for a largest range of a_T.

For the Gaussian case, we assume in addition that $\xi(t)$ is a modification of the Wiener process which has continuous trajectories w.p. 1.

We concern with the processes having non-negative linear drifts, that is the c.f. of $\xi(1)$ is

$$\psi(t) = \exp\left\{i\mu t - c|t|^\alpha \left(1 + i\frac{t}{|t|}\mathrm{tg}\frac{\pi}{2}\alpha\right)\right\}, \tag{4.11}$$

where $c = \cos(\pi(2-\alpha)/2)/\alpha$, $1 < \alpha \leq 2$ and $\mu \geq 0$. Since $\xi(1) - \mu$ has the c.f. (2.2), we have

$$\gamma(x) = \mu + \left(\frac{\alpha x}{\alpha - 1}\right)^{(\alpha-1)/\alpha}$$

for all $x \geq 0$. Taking this into account, we arrive at the next result.

Theorem 4.5. *If $\xi(1)$ has c.f. (4.11) with $\mu \geq 0$, then relation (4.9) holds with*

$$b_T = \mu a_T + \lambda^{-\lambda} a_T^{1/\alpha} \left(\log \frac{T}{a_T} + \log\log T\right)^\lambda, \tag{4.12}$$

where $\lambda = (\alpha - 1)/\alpha$.

If, in addition, $\log\log T = o(\log(T/a_T))$ or $a_T = T$ and $\mu > 0$, then (4.10) holds with b_T from (4.12).

Theorem 4.5 follows from Theorem 4.4 and the formula for b_T. Condition (4.2) may be easily checked. Condition (4.4) for the stable laws holds obviously. One can check inequality (4.7) with an application of Remark 4.1 and Lemma 2.1. For $a_T = O(\log T)$ and $\mu > 0$, we get Theorem 4.5 from the result for $\mu = 0$ and the properties of $\gamma(x)$. For $a_T \to \infty$ and $\mu > 0$, Theorem 4.5 follows from Theorem 4.13 which will be proved below.

Theorem 4.5 includes the SLLN, the LIL, the Erdős–Rényi law and the Csörgő–Révész laws for the Wiener process and the stable processes without positive jumps.

If $\alpha = 2$ and $\mu = 0$, then $\xi(t)$ is the standard Wiener process and

$$b_T = \sqrt{2a_T\left(\log\frac{T}{a_T} + \log\log T\right)}. \tag{4.13}$$

For $a_T = T$, we have the LIL with $b_T = \sqrt{2T\log\log T}$. If $\alpha < 2$ and $\mu = 0$, then $\xi(t)$ is a stable process without positive jumps and the LIL holds with $b_T = \lambda^{-\lambda}T^{1/\alpha}(\log\log T)^{\lambda}$.

Note that for $\mu = 0$, Theorem 4.5 also yields the Erdős–Rényi law and the Mason's extension of this law when $a_T = c\log T$ and $a_T = o(\log T)$ correspondingly. Remember, that $c_0 = 0$ in this case. If $\mu > 0$ and $a_T = O(\log T)$, the situation is the same.

When $\mu > 0$ and $a_T/\log T \to \infty$, then μa_T is the dominating part of the norming in (4.12). We consider this case in Theorem 4.13 below. From there, the SLLN follows, in particular. We included the SLLN for the Wiener process and the stable processes without positive jumps ($\alpha \in (1, 2)$) in Theorem 4.5 as the case $a_T = T$ for $\mu > 0$. Note that the latter yields the SLLN for arbitrary μ.

One realizes that the behaviours of the Wiener and the stable process without positive jumps are very similar. Nevertheless, this similarity of is not absolute. To see that, we have the result which holds for the Wiener process only.

Theorem 4.6. *If $\xi(t)$ is the standard Wiener process, then the conclusion of Theorem 4.5 holds true provided the increments of the process are replaced by their moduli in W_n, U_n, R_n and T_n.*

4.3 Applications of the Universal Strong Laws

In this section, we discuss various corollaries of Theorem 4.4 for arbitrary processes with independent increments satisfying the conditions of Section 4.1. In general case, we can not write one simple formula of norming sequence for all a_T as in the previous section. Hence, we further consider small and large increments separately.

Assume first that $a_T = O(\log T)$ and start with the Erdős–Rényi and Spepp laws for processes with independent increments.

Theorem 4.7. *If $h_0 > 0$ and $a_T = c\log T$, $c > c_0$, then*

$$\lim\frac{W_T}{a_T} = \lim\frac{U_T}{a_T} = \limsup\frac{R_T}{a_T} = \limsup\frac{Q_T}{a_T} = \gamma\left(\frac{1}{c}\right) \quad a.s.$$

Theorem 4.7 follows from Theorem 4.4 and the formula for b_T. Condition (4.2) may be easily checked. Condition (4.4) for the stable laws holds obviously. In Theorem 4.7, condition (4.4) follows from Remark 4.1 and the weak law of large numbers for i.i.d. random variables. One can check inequality (4.7) with an application of Remark 4.1 and Lemma 2.1.

Suppose now that $a_T = o(\log T)$. In this case, we have the following analogue of Mason's extension for the Erdős–Rényi law.

Theorem 4.8. *Assume that $h_0 > 0$ and $a_T = o(\log T)$. If the conditions of Theorems 4.1 and 4.2 are satisfied, then relation (4.9) holds with $b_T = a_T\gamma(\log T/a_T)$. If the conditions of Theorems 4.1 and 4.3 are satisfied, then (4.10) holds with the same b_T.*

Theorem 4.8 follows from Theorem 4.4, the formula for b_T and the concavity of the function $\gamma(x)$.

We stated the Erdős–Rényi and Shepp laws for processes with independent increments for $c > c_0$ only. For $c \leq c_0$, results are similar to that for sums of i.i.d. random variables. We do not state them here.

Assume now that $a_T/\log T \to \infty$. The behaviour of large increments depends on moment conditions on the right-hand tail of $\xi(1)$ in the same manner as for sums of i.i.d. random variables.

It is clear that results for centered ($\mu = 0$) and non-centered increments are quite different. For example, if $a_T = T$, then we have the LIL for centered increments and the SLLN for non-centered ones. We deal with these cases separately in the sequel.

Consider first centered processes. We have Csörgő–Révész laws in this case. Put $\xi^+(1) = \max\{\xi(1), 0\}$.

Theorem 4.9. *Assume that $\mu = 0$, $E\xi^2(1) = 1$ and one the following conditions holds:*

1) $h_0 > 0$ and $a_T/\log T \to \infty$;

2) $Ee^{(\xi^+(1))^\beta} < \infty$ for some $\beta \in (0,1)$ and $a_T \geq C(\log T)^{2/\beta-1}$, where $C = C(\beta)$ is an absolute positive constant;

3) $E\xi^+(1)^p < \infty$ for some $p > 2$ and $a_T \geq CT^{2/p}/\log T$ where C is an arbitrary positive constant.

Then (4.9) holds with b_T from (4.13). If, in addition, $\log\log T = o(\log(T/a_T))$, *then (4.10) holds with b_T from (4.13).*

Theorem 4.9 yields the following result on the LIL for increments of processes with independent increments.

Corollary 4.1. *Assume that* $\mu = 0$, $E\xi^2(1) = 1$ *and* $E\xi^+(1)^p < \infty$ *for some* $p > 2$. *If* $\log(T/a_T)/\log\log T \to c$, *then*

$$\limsup \frac{W_T}{\sqrt{2a_T \log\log T}} = \limsup \frac{U_T}{\sqrt{2a_T \log\log T}} = \sqrt{1+c} \quad a.s.$$

If $a_T = T$, then from Corollary 4.1, we obtain the LIL for processes with independent increments.

Turn to the case $E\xi^2(1) = \infty$. Put $F(x) = P(\xi(1) < x)$.

We start with the case of non-normal attraction to the normal law. We then have the following result.

Theorem 4.10. *Assume that* $\mu = 0$, $F \in D(2)$ *and one of conditions* $1) - 3)$ *of Theorem 4.9 is satisfied. Then (4.9) holds with*

$$b_T = a_T \hat{m}(\hat{f}^{-1}(d_T)), \tag{4.14}$$

where

$$\hat{m}(h) = hG\left(\frac{1}{h}\right), \quad \hat{f}(h) = \frac{h^2}{2} G\left(\frac{1}{h}\right), \quad G(x) = \int_{-x}^{0} u^2 dF(u), \quad x > 0.$$

($G(x) \in SV_\infty$.) If, in addition, $\log\log T = o(\log(T/a_T))$, *then (4.10) holds with* b_T *from (4.14).*

If $\log(T/a_T)/\log\log T \to c$, then Theorem 4.10 implies the LIL for increments of processes under consideration. For $a_T = T$, we arrive at the LIL. The result is similar to that of Corollary 4.1, but a slowly varying multiplier will appear in the normalizing function.

Turn to the case $\alpha \in (1, 2)$. For domains of normal attraction of the stable laws, the result is as follows.

Theorem 4.11. *Assume that* $\mu = 0$, $F \in DN(\alpha)$ *for some* $\alpha \in (1, 2)$ *and one of the following conditions holds:*

1) $h_0 > 0$ *and* $a_T/\log T \to \infty$;

2) $Ee^{(\xi^+(1))^\beta} < \infty$ *for some* $\beta \in (0, 1)$ *and* $a_T \geq C(\log T)^{\alpha/\beta - \alpha - 1}$, *where* $C = C(\alpha, \beta)$ *is an absolute positive constant;*

3) $E\xi^+(1)^p < \infty$ *for some* $p > 2$, $a_T \geq CT^{\alpha/p}/(\log T)^{\alpha-1}$ *where* C *is an arbitrary positive constant.*

Then (4.9) holds with b_T *from (4.12). If, in addition,* $\log\log T = o(\log(T/a_T))$, *then (4.10) holds with* b_T *from (4.12).*

Theorem 4.11 implies the next result on the LIL for increments of processes with independent increments.

Corollary 4.2. *Assume that* $\mu = 0$, $E\xi^2(1) = 1$ *and* $E\xi^+(1)^p < \infty$ *for some* $p > 2$. *If* $\log(T/a_T)/\log\log T \to c$, *then*

$$\limsup \frac{W_T}{\lambda^{-\lambda} a_T^{1/\alpha} (\log\log T)^\lambda} = \limsup \frac{U_T}{\lambda^{-\lambda} a_T^{1/\alpha} (\log\log T)^\lambda} = (1+c)^\lambda \quad a.s.$$

For $a_T = T$, then Corollary 4.2 turns to the LIL for processes with independent increments.

For domains of non-normal attraction, we have the following result.

Theorem 4.12. *Assume that* $\mu = 0$ *and* $F \in D(\alpha)$ *for some* $\alpha \in (1,2)$. *Put*

$$b_T = a_T \hat{m}(\hat{f}^{-1}(d_T)), \tag{4.15}$$

where

$$\hat{m}(h) = \frac{\alpha\Gamma(2-\alpha)}{\alpha - 1} h^{\alpha-1} G\left(\frac{1}{h}\right), \quad \hat{f}(h) = \Gamma(2-\alpha) h^\alpha G\left(\frac{1}{h}\right),$$

and $G(x) = x^\alpha F(-x)$, $x > 0$. *(*$G(x) \in SV_\infty$.*)*

Assume that one the following conditions holds:

1) $h_0 > 0$ *and* $a_T/\log T \to \infty$;

2) $Ee^{(\xi^+(1))^\beta} < \infty$ *for some* $\beta \in (0,1)$ *and* $b_T \geq C(\log T)^{1/\beta}$, *where* $C = C(\alpha, \beta)$ *is an absolute positive constant;*

3) $E\xi^+(1)^p < \infty$ *for some* $p > \alpha$ *and* $b_T \geq CT^{1/p}$, *where* C *is an arbitrary positive constant.*

Then (4.9) holds with b_T *from (4.15). If, in addition,* $\log\log T = o(\log(T/a_T))$, *then (4.10) holds with* b_T *from (4.15).*

For $\log(T/a_T)/\log\log T \to c$, Theorem 4.10 yields the LIL for increments and the LIL for processes with independent increments when $a_T = T$. The result is similar to that of Corollary 4.2, but a slowly varying multiplier will appear in the normalizing function.

Proofs of Theorems 4.9–4.12 consist of choices of y_T (if $h_0 = 0$), calculations of b_T and verifications of corresponding conditions for b_T. They are the same as for sums of i.i.d. random variables. Indeed, $\xi(n) = \sum_{k=1}^{n} (\xi(k) - \xi(k-1))$, where $\{\xi(k) - \xi(k-1)\}$ is a sequence of i.i.d. random variables. Remember that we may write the conditions of Theorem 4.4 for natural times. Thus, we omit details.

We finally deal with the case $\mu > 0$.

Theorem 4.13. *Assume that the conditions of one of Theorems 4.9–4.12 are satisfied and $\mu > 0$ (instead of $\mu = 0$). Then (4.9) holds with $b_T = \mu a_T$. If, in addition, $\log\log T = o(\log(T/a_T))$ or $a_T = T$, then (4.10) holds with the same b_T.*

Theorem 4.13 for $a_T = T$ implies the SLLN for processes with $\mu > 0$. Applying of this result to the process $\xi(t) - \mu t + t$, we get the SLLN for $\mu \leq 0$ as well.

Proof. We have $b_n \sim \mu a_n$ as $n \to \infty$ (see the proof of Theorem 3.16).

If $n \leq T < n+1$, then $b_n \leq b_T \leq (1+\varepsilon)\mu a_{n+1} \leq (1+\varepsilon)\mu a_T(n+1)/T$ for every $\varepsilon > 0$ and all sufficiently large n. By Theorem 4.4, we get the result besides the SLLN in the case $a_T = T$ for all T.

Assume now that $a_T = T$. Then (see the proof of Theorem 3.16) $\xi(n)/n \to \mu$ a.s.

Put

$$A_n = \left\{ \sup_{n-1 \leq t < n} |\xi(t) - \xi(n)| \geq 2\varepsilon n \right\}$$

for $n \in \mathbf{N}$ and $\varepsilon > 0$. For all n, we have

$$\begin{aligned} P(A_n) &\leq P\left(\sup_{0 \leq t \leq 1} |\xi(t)| \geq 2\varepsilon n \right) \\ &= P\left(\sup_{0 \leq t \leq 1} \xi(t) \geq 2\varepsilon n \right) + P\left(\sup_{0 \leq t \leq 1} (-\xi(t)) \geq 2\varepsilon n \right). \end{aligned}$$

For all $t \in [0,1]$, we get

$$P(\xi(t) \geq -c) \geq P\left(\sup_{0 \leq t \leq 1} \xi(t) \geq -c \right) \geq 0.5$$

provided c is large enough. By Lemma 4.1, we obtain

$$P\left(\sup_{0 \leq t \leq 1} \xi(t) \geq 2\varepsilon n \right) \leq 4P(\xi(1) \geq 2\varepsilon n - 2c) \leq 4P(\xi(1) \geq \varepsilon n)$$

for all sufficiently large n. Note that Lemma 4.1 holds for $-\xi(t)$ as well. In the same way, we prove that

$$P\left(\sup_{0 \leq t \leq 1} (-\xi(t)) \geq 2\varepsilon n \right) \leq 4P(-\xi(1) \geq \varepsilon n)$$

for all sufficiently large n.

It follows that

$$P(A_n) \le 4P(|\xi(1)| \ge \varepsilon n)$$

for all sufficiently large n. Since $E|\xi(1)| < \infty$, the series $\sum_n P(A_n)$ converges. By the Borel–Cantelli lemma, we have $P(A_n \text{ i.o.}) = 0$ and the result follows. □

In the proof of Theorem 4.13, we applied Theorem 3.16 that is the corollary of the universal strong laws.

Previous results can not hold for moduli of increments of all considered processes. In this case, $\xi(1)$ and $-\xi(1)$ have to satisfy the same moment assumptions of the theorems discussed above. The latter is only possible for $E\xi(1) = 0$ and $E\xi(1)^2 < \infty$.

Turn now to results for moduli of increments. Put

$$W_T^\star = \sup_{0 \le t \le T - a_T} \sup_{0 \le s \le a_T} |\xi(t+s) - \xi(t)|,$$
$$U_T^\star = \sup_{0 \le t \le T - a_T} |\xi(t+a_T) - \xi(t)|,$$
$$Q_T^\star = |\xi(T+a_T) - \xi(T)|, \quad R_T^\star = |\xi(T) - \xi(T-a_T)|.$$

From Theorem 4.9 and the equality $|X| = \max\{X, -X\}$, we obtain the following result.

Theorem 4.14. *Assume that* $\mu = 0$, $E\xi^2(1) = 1$ *and one the following conditions holds:*

1) $\mathrm{E}e^{h^\star|\xi(1)|} < \infty$ *for some* $h^\star > 0$ *and* $a_T/\log T \to \infty$*;*

2) $\mathrm{E}e^{|\xi(1)|^\beta} < \infty$ *for some* $\beta \in (0,1)$ *and* $a_T \ge C(\log T)^{2/\beta - 1}$, *where* $C = C(\beta)$ *is an absolute positive constant;*

3) $E|\xi(1)|^p < \infty$ *for some* $p > 2$ *and* $a_T \ge CT^{2/p}/\log T$ *where* C *is an arbitrary positive constant.*

Then the relation

$$\limsup \frac{W_T^\star}{b_T} = \limsup \frac{U_T^\star}{b_T} = \limsup \frac{R_T^\star}{b_T} = \limsup \frac{Q_T^\star}{b_T} = 1 \quad a.s. \tag{4.16}$$

holds with b_T *from (4.13). If, in addition,* $\log\log T = o(\log(T/a_T))$, *then the relation*

$$\lim \frac{W_T^\star}{b_T} = \lim \frac{U_T^\star}{b_T} = 1 \quad a.s. \tag{4.17}$$

holds with b_T *from (4.13).*

Theorem 4.14 is the Csőrgö–Révész law for moduli of increments of processes with independent increments.

Theorem 4.14 implies the following result on the LIL for moduli of increments of processes with independent increments.

Corollary 4.3. *Assume that* $\mu = 0$, $E\xi^2(1) = 1$ *and* $E|\xi(1)|^p < \infty$ *for some* $p > 2$. *If* $\log(T/a_T)/\log\log T \to c$, *then*

$$\limsup \frac{W_T^\star}{\sqrt{2a_T \log\log T}} = \limsup \frac{U_T^\star}{\sqrt{2a_T \log\log T}} = \sqrt{1+c} \quad a.s.$$

If $a_T = T$, then from Corollary 4.3, we obtain the LIL for moduli of processes with independent increments.

For SLLN, we get the next result.

Theorem 4.15. *If the conditions of Theorem 4.14 hold, then*

$$\lim \frac{W_T^\star}{a_T} = \lim \frac{U_T^\star}{a_T} = \lim \frac{R_T^\star}{a_T} = \lim \frac{Q_T^\star}{a_T} = 0 \quad a.s.$$

In view of non-negativity of the considered functionals, Theorem 4.15 follows from Theorem 4.14 and the relation $b_n = o(a_n)$ which is a result of $a_n/\log n \to \infty$.

4.4 Compound Poisson Processes

Now we describe an important class of processes with independent increments which the results of previous section hold for.

Let $\{X_k\}$ be a sequence of i.i.d. random variables and $\nu(t)$ be a Poisson process independent with X's. Then the process

$$\xi(t) = \sum_{k=1}^{\nu(t)} X_k, \quad \xi(0) = 0,$$

is called the Compound Poisson process.

Such processes play an important role in the probability theory and its applications. In financial and actuarial mathematics, they are models for claim processes. For example, if $\nu(t)$ is the number of claims up to time t and X_i is the amount of i-th claim, then $\xi(t)$ is the aggregate amount of claims up to time t over a portfolio of insurance policies of the same type.

It is not difficult to check that the c.f. of $\xi(t)$ is

$$\Psi_t(u) = Ee^{iu\xi(t)} = e^{\nu t(g(u)-1)},$$

where $g(u) = Ee^{iuX_1}$ and ν is a parameter of the Poisson process $\nu(t)$. It implies that $\xi(t)$ is a process with independent increments.

Moreover, the formula for $\Psi_1(u)$ allows to calculate moments of $\xi(1)$ provided they exist. In particular, calculating the derivatives of $\Psi_1(u)$ at zero, we get

$$\mu = E\xi(1) = \nu EX_1, \quad E\xi(1)^2 = \nu EX_1^2 + \nu^2 (EX_1)^2, \quad D\xi(1) = \nu EX_1^2.$$

The latter means that $\xi(t)$ may be centered and non-centered. Remember that the results of previous section crucially depends on that.

Further, it is not difficult to check that the m.g.f. of $\xi(1)$ satisfies the relation

$$\varphi(h) = e^{\nu\left(Ee^{hX_1} - 1\right)}$$

for every h such that $Ee^{hX_1} < \infty$.

To apply the results of the previous section for $E\xi(1)^2 = \infty$, we need conditions sufficient for $F \in D(\alpha)$, where $F(x)$ is the d.f. of $\xi(1)$. It turns out that the d.f. of $\xi(1)$ belongs to the domain of attraction of considered stable laws provided the d.f. of X_1 belongs to the same domain of attraction. To check this, we prove the next result.

Lemma 4.2. *If $EX_1 = 0$ and $P(X_1 < x) \in DN(\alpha)$ or $P(X_1 < x) \in DN(\alpha)$ with $\alpha \in (1, 2]$, then $F(x) \in DN(\alpha)$ or $F(x) \in DN(\alpha)$ correspondingly.*

Proof. If $EX_1 = 0$ and the d.f. of X_1 is from the domain of attraction of the stable law with the c.f. $\psi(t)$ from (2.2), then for every fixed $u \in \mathbb{R}$, we have

$$g^n\left(\frac{u}{B_n}\right) \to \psi(u),$$

where $\{B_n\}$ is an appropriate sequence of norming constant with $B_n \to \infty$ as $n \to \infty$. It turns out that

$$n\left(g^n\left(\frac{u}{B_n}\right) - 1\right) = n \log\left(g^n\left(\frac{u}{B_n}\right)\right) + o(1) = \log \psi(u) + o(1)$$

as $n \to \infty$. It follows that

$$\Psi_1^n\left(\frac{u}{B_n}\right) = \exp\left\{n\left(g^n\left(\frac{u}{B_n}\right) - 1\right)\right\} \to \psi(u)$$

as $n \to \infty$. The latter means that $\xi(1)$ is from the domain of attraction of the same stable law as X_1. □

It follows that all results of the previous section hold under appropriate assumptions on the distribution of X_1.

If $P(X_1 = 1) = 1$ in the definition of the Compound Poisson process, then $\xi(t)$ tuns to the Poisson process. It is not centered and, therefore, we have the SLLN (Theorem 4.13) and the Erdős–Rényi laws (Theorems 4.7 and 4.8). But for $\xi(t)$ centered at mean, we will have the Csörgő–Révész laws (Theorem 4.9) and LIL (Corollary 4.1). Results for moduli of increments holds as well.

4.5 Bibliographical Notes

Results for increments of Wiener process are proved by [Book and Shore (1978)], [Csáki and Révész (1979)], [Csörgő and Révész (1979)], [Hanson and Russo (1983)] and [Ortega and Wschebor (1984)]. Increments of stable processes with jumps of one sign have been studied by [Zinchenko (1987)].

LIL for asymmetric stable processes may be found in [Fristedt (1964)], [Brieman (1968)], [Millar (1972)] and [Mijnheer (1974)].

The universal strong laws for homogeneous processes with independent increments are obtained in [Frolov (2003a, 2004c)].

Many results on increments of various stochastic processes (processes related with the Wiener process, empirical processes, local times) and references may be found in monographs [Csörgő and Révész (1981)] and [Révész (1990)].

In this book, we do not deal with random fields, but the situation is similar to those for sums of i.i.d. random variables and processes with independent increments.

The behaviour of increments of random fields is investigated in [Steinebach (1983)], [Deheuvels (1985)], [Pfuhl and Steinebach (1988)], [Scherbakova (2003, 2004)], [Frolov (2003c, 2002d)]. Asymptotics of increments of Gaussian and stable random fields may be derived from those of corresponding multiparameter processes. Results for multiparameter Gaussian processes are obtained in [Chan (1976)], [Csörgő and Révész (1978, 1981)], [Lin *et al.* (2001)], [Choi and Kôno (1999)]. Results for stable processes are proved in [Zinchenko (1994)]. The universal strong laws for random fields may be found in [Frolov (2002d, 2003c)].

Chapter 5

Strong Limit Theorems for Renewal Processes

Abstract. Universal strong laws are derived for renewal processes. Results include the SLLN, the LIL, the Erdős–Rényi–Shepp laws and the Csörgő–Révész laws. Poisson processes are partial cases.

5.1 The universal Strong Laws for Renewal Processes

Let $Y, Y_1, Y_2, \ldots$ be a sequence of i.i.d. non-degenerate random variables such that $\operatorname{ess\,inf} Y = 0$ and $\mu = EY < \infty$. Put $R_0 = 0$,

$$R_n = Y_1 + Y_2 + \cdots + Y_n$$

for $n \in \mathbf{N}$ and define the renewal process by

$$N(t) = \max\{n \geq 0 : R_n \leq t\}, \quad t \geq 0.$$

Note that if Y has the exponential distribution with the density

$$p(x) = \frac{1}{\mu} e^{-x/\mu} I_{[0,+\infty)}(x),$$

then $N(t)$ is the Poisson process and $EN(t) = t/\mu$.

Let a_T be a non-decreasing, positive function such that $a_T \leq T$, T/a_T is a non-decreasing function and $a_T \to \infty$. Put

$$W_T = \sup_{0 \leq t \leq T - a_T} \sup_{0 \leq s \leq a_T} \Big(N(t+s) - N(t) - \frac{s}{\mu} \Big),$$

$$u_T = \sup_{0 \leq t \leq T - a_T} (N(t + a_T) - N(t)), \quad U_T = u_T - \frac{a_T}{\mu}.$$

Note that $u_T = N(T)$ for $a_T = T$. The maximum U_T is a centered variant of u_T. The centering function corresponds to that from the SLLN for renewal processes. A non-centered analogue for W_T coincides with u_T since $N(T)$ is non-decreasing.

We find necessary and/or sufficient conditions for

$$\limsup \frac{W_T}{b_T} = 1 \quad \text{a.s.},$$

where b_T is a non-decreasing positive function. Sometimes, one can replace lim sup by lim. The behaviour of U_T and u_T is discussed as well.

Writing lim sup, lim inf, lim, O, o, $\to$, we assume that $T \to \infty$, if it is not pointed otherwise.

We follow the pattern of the previous chapter. We first find a formula for the norming function b_T. Further, we state the universal strong laws for increments of renewal processes. Finally, we derive corollaries which include the Erdős–Rényi and Shepp laws, the Csörgő–Révész laws, the LIL and the SLLN. We consider renewal times from domains of attraction of completely asymmetric stable laws with the exponent $\alpha \in (1, 2]$. Renewal times with finite variations are partial cases.

Put $X = \mu - Y$ and $F(x) = P(X < x)$. The random variable X is bounded from above and $\omega = \mu$. It is clear that either $F \in K_1$, or $F \in K_2$. We use the functions of the LDT $\varphi(h) = Ee^{hX}$, $m(h)$, $\sigma^2(h)$, $f(h)$, $\zeta(z)$ and $\gamma(x)$ from Chapter 2.

Let c_T be a non-decreasing function such that

$$c_T\Big(\mu - \gamma\Big(\frac{\log(T/c_T) + \log\log T}{c_T}\Big)\Big) = a_T \tag{5.1}$$

and T/c_T is non-decreasing for all sufficiently large T. If $F \in K_2$, then we assume in addition that $c_T/\log T > c_0$.

It is not difficult to check that by the properties of $\gamma(x)$, relation (5.1) determines the function c_T from a_T and, moreover, $c_T > a_T/\mu$ for all T. Put

$$b_T = \begin{cases} c_T - (a_T/\mu), & \text{for} \quad U_T \text{ and } W_T, \\ c_T, & \text{for} \quad u_T. \end{cases} \tag{5.2}$$

It is clear that $b_T > 0$ for all T. Since $a_T \to \infty$, we have $b_T \to \infty$. For u_T, this is obvious. Otherwise, it easily follows from the properties of $\gamma(x)$ and

$$\mu b_T = c_T \gamma\Big(\frac{\log(T/c_T) + \log\log T}{c_T}\Big). \tag{5.3}$$

The latter is another form of relation (5.1).

For $F \in K_2$ and $c_T/\log T \to c_0$, relations (5.1) and (5.3) yield that $a_T = o(c_T)$ and $b_T \sim c_0 \log T$.

We assume in what follows that b_T is equivalent to a non-decreasing function.

We start with the next result.

Theorem 5.1. *Assume that*

$$\limsup_{\theta \searrow 1} \limsup_{T\to\infty} \frac{b_{\theta T}}{b_T} = 1$$

and for every $\varepsilon > 0$ *there exists* $q \in (0,1)$ *such that*

$$P(S_{[(1+\varepsilon)(1+3\varepsilon)c_T]} \geq -\varepsilon \mu b_T) \geq q \tag{5.4}$$

for all sufficiently large T. *Here and in the sequel,* $S_n = n\mu - R_n$ *for all* $n \in \mathbf{N}$ *and* $[\cdot]$ *is the integer part of the number in brackets.*

Then

$$\limsup \frac{W_T}{b_T} \leq 1 \quad a.s.$$

One can replace W_T *by* u_T *in the last relation. For* $c_T = O(\log T)$, *one can omit condition (5.4).*

Remember that replacing W_T by u_T, one changes b_T in view of (5.2).

We first establish a duality between increments of renewal processes and sums S_n. It is a basic tool of our proofs below.

Lemma 5.1. *The following relations hold:*

$$\{W_T \geq x+1\} \subset \left\{ \max_{[x]\leq j\leq [x+\mu^{-1}a_T]+1} \; \max_{0\leq k\leq N(T)-j} (S_{k+j} - S_k) \geq \mu x \right\}, \tag{5.5}$$

$$\{U_T < x-2\} \subset \left\{ \max_{0\leq k\leq N(T)-[x+\mu^{-1}a_T]+1} (S_{k+[x+\mu^{-1}a_T]} - S_k) < \mu x \right\}, \tag{5.6}$$

$$\{U_T < x-2\} \subset \{S_{[x+\mu^{-1}a_T]} < \mu x\}, \tag{5.7}$$

$$\{u_T < x-2\} \subset \left\{ \min_{0\leq k\leq N(T)-[x]+1} (R_{k+[x]} - R_k) \geq a_T \right\}. \tag{5.8}$$

Proof. If $W_T \geq x+1$, then $N(t+s) - (x+1+s\mu^{-1}) \geq N(t)$ for some $s \in [0, a_T]$ and $t \in [0, T-a_T]$. Since R_n is non-decreasing, we have

$$R_{[N(t+s)-(x+1+s\mu^{-1})]+1} \geq R_{N(t)+1} > t \geq R_{N(t+s)} - s$$

by the definition of $N(t)$. The last inequality yields that

$$S_{N(t+s)} - S_{[N(t+s)-(x+s\mu^{-1})]} \geq x\mu.$$

Put $k = [N(t+s)-(x+s\mu^{-1})]$ and $j = (x+s\mu^{-1}) + \{N(t+s)-(x+s\mu^{-1})\}$, where $\{\cdot\}$ is a fractional part of the number in brackets. Then $N(t+s) = k+j$, $[x] \leq j \leq [x+a_T\mu^{-1}]+1$ and $0 \leq k \leq N(T)-j$. Hence, the inequality

$$S_{k+j} - S_k \geq x\mu$$

holds. Relation (5.5) follows.

If $U_T < x - 2$, then $N(t + a_T) - N(t) < [x + a_T\mu^{-1}] - 1$ for all $t \in [0, T - a_T]$. Since R_n is non-decreasing, we have

$$R_{N(t+a_T)-[x+a_T\mu^{-1}]+1} < R_{N(t)} \leq t < R_{N(t+a_T)+1} - a_T$$

by the definition of $N(t)$. It follows from the last inequality that

$$S_{N(t+a_T)+1} - S_{N(t+a_T)-[x+a_T\mu^{-1}]+1} < \mu x$$

for all $t \in [0, T - a_T]$. Putting $k = N(t + a_T) - [x + a_T\mu^{-1}] + 1$ implies that for all k with $0 \leq k \leq N(T) - [x + a_T\mu^{-1}] + 1$, the inequality

$$S_{k+[x+a_T\mu^{-1}]} - S_k < x\mu$$

holds. This yields relations (5.6) and (5.7).

Relation (5.8) may be proved in the same way as (5.6). We omit details. □

We turn to the proof of Theorem 5.1.

Proof. Take $\varepsilon > 0$ and $\delta > 0$. Put $K = [(1 + \delta)T\mu^{-1}]$ and $J = [(1 + 3\varepsilon)b_T + a_T\mu^{-1}]$. Since $b_T \to \infty$, an application of (5.5) with $x = (1 + 3\varepsilon)b_T$ yields

$$\begin{aligned}
&P(W_T \geq (1 + 4\varepsilon)b_T) \leq P(W_T \geq (1 + 3\varepsilon)(b_T + 1)) \\
&\leq P\left(\max_{1\leq j\leq J+1}\ \max_{0\leq k\leq N(T)-j}(S_{k+j} - S_k) \geq (1 + 3\varepsilon)\mu b_T\right) \\
&\leq P\left(\max_{1\leq j\leq J+1}\ \max_{0\leq k\leq K}(S_{k+j} - S_k) \geq (1 + 3\varepsilon)\mu b_T\right) + P(N(T) \geq K) \\
&= P_T + Q_T
\end{aligned}$$

for all sufficiently large T.

We first check that every $\varepsilon \in (0, \varepsilon_0)$ there exists $\tau > 0$ such that

$$P_T \leq (\log T)^{-(1+\tau)} \tag{5.9}$$

for all sufficiently large T, where ε_0 is an absolute constant.

Put

$$d_T = \log\frac{T}{c_T} + \log\log T. \tag{5.10}$$

Suppose that $\varepsilon c_T \leq \log T$. The concavity of $\gamma(x)$ and (5.3) imply that

$$(1 + 3\varepsilon)\mu b_T \geq j\gamma\left((1 + 3\varepsilon)\frac{d_T}{j}\right)$$

for $1 \le j \le J_1 = [(1+3\varepsilon)c_T]$. Taking into account Lemma 2.2 and the definition of $\gamma(x)$, we have

$$P_T \le K \sum_{j=1}^{J+1} P(S_j \ge (1+3\varepsilon)\mu b_T) \le K \sum_{j=1}^{J_1} P\left(S_j \ge j\gamma\left((1+3\varepsilon)\frac{d_T}{j}\right)\right)$$

$$\le K \sum_{j=1}^{J_1} \mathrm{e}^{-(1+3\varepsilon)d_T} \le K J_1 T^{-(1+3\varepsilon)} c_T^{1-3\varepsilon} (\log T)^{-(1+3\varepsilon)}$$

for all sufficiently large T. This implies (5.9).

Suppose now that $\varepsilon c_T > \log T$. Put $m = [\varepsilon J]$, $L = [K/m] + 1$ and $J_2 = J + m + 1$. Note that $J_2 \sim (1+\varepsilon)(c_T + 3\varepsilon b_T)$. By (5.3), we get

$$b_T = \mu^{-1} c_T \gamma\left(\frac{d_T}{c_T}\right) \le \mu^{-1} c_T \gamma(2\varepsilon)$$

for all sufficiently large T. Hence, for all $\varepsilon < \varepsilon_0$, the inequalities

$$(1+\varepsilon)c_T \ge (1+\varepsilon)\{(1+0.1\varepsilon)(1+\varepsilon)(1+3\varepsilon\mu^{-1}\gamma(2\varepsilon))\}^{-1} J_2 \ge J_2$$

hold for all sufficiently large T.

By Lemma 1.1, condition (5.4) and (5.3), we have

$$P_T \le \sum_{l=1}^{L} P\left(\max_{1\le j\le J+1} \max_{(l-1)m\le k\le lm} (S_{k+j} - S_k) \ge (1+3\varepsilon)\mu b_T\right)$$

$$\le LP\left(\max_{1\le j\le J+1} \max_{0\le k\le m} (S_{k+j} - S_k) \ge (1+3\varepsilon)\mu b_T\right)$$

$$\le \frac{L}{q^2} P(S_{J_2} \ge (1+\varepsilon)\mu b_T) \le \frac{K+m}{mq^2} P\left(S_{J_2} \ge J_2 \gamma\left(\frac{d_T}{c_T}\right)\right)$$

for all sufficiently large T. This bound, Lemma 2.2 and the definition of $\gamma(x)$ yield (5.9).

We now examine the rate of decreasing for Q_T. We have

$$Q_T = P\Big(R_K \le \frac{K\mu}{1+\delta}\Big) = P\Big(S_K \ge \frac{K\mu\delta}{1+\delta}\Big) \le \mathrm{e}^{-K\zeta(\delta\mu/(1+\delta)} = \mathrm{e}^{-\rho T(1+o(1))}.$$

Take $\theta > 1$. Put $T_k = \theta^k$ for all k. By (5.9), we get

$$\sum_{k=1}^{\infty} P(W_{T_k} \ge (1+4\varepsilon)b_{T_k}) \le \sum_{k=1}^{\infty} P_{T_k} + \sum_{k=1}^{\infty} Q_{T_k} < \infty.$$

The Borel–Cantelli lemma implies that

$$\limsup_{k\to\infty} \frac{W_{T_k}}{b_{T_k}} \le 1 + 4\varepsilon \quad \text{a.s.} \tag{5.11}$$

Further, $b_{T_k} \le b_{T_{k+1}} \le b_{\theta^3 T_k} \le (1+\tau)b_{T_k}$ for all $\tau > 0$, θ closed to 1 and large k. Since W_T is non-decreasing, we have

$$\frac{W_T}{b_T} \le (1+\tau)\frac{W_{T_{k+1}}}{b_{T_{k+1}}}$$

for $T_k \le T \le T_{k+1}$. The last inequalities and (5.11) yield

$$\limsup \frac{W_T}{b_T} \le (1+\tau)(1+4\varepsilon) \quad \text{a.s.}$$

Passing to the limit as $\tau \to 0$ and $\varepsilon \to 0$, we arrive at the result for W_n. Taking into account that

$$\{u_T \ge (1+\varepsilon)c_T\} \subset \left\{U_T \ge (1+\varepsilon)\left(c_T - \frac{a_T}{\mu}\right)\right\}$$
$$\subset \left\{W_T \ge (1+\varepsilon)\left(c_T - \frac{a_T}{\mu}\right)\right\},$$

we obtain the result for u_T. □

Now we turn to lower bounds.

Theorem 5.2. *Assume that* $\log T/c_T < \min\{1/c_0 - \varepsilon_0, 1/\varepsilon_0\}$ *for all sufficiently large* T *and some* $\varepsilon_0 > 0$. *If* $a_T/T \to 1$, *then suppose additionally that condition (5.4) is satisfied. Then*

$$\limsup \frac{U_T}{b_T} \ge 1 \quad a.s.$$

One can replace U_T *by* u_T *in the last relation.*

Proof. Take $\varepsilon > 0$ and $\delta > 0$. Put $K = K(T) = [(1-\delta)T\mu^{-1}]$, $J = J(T) = [(1-\varepsilon)b_T + a_T\mu^{-1}]$.

Assume first that $a_T/T < 1-\rho$ for all sufficiently large T. By (5.6), we have

$$\{U_T \ge (1-2\varepsilon)b_T\} \supset \{U_T \ge (1-\varepsilon)b_T - 2\}$$
$$\supset \left\{\max_{0\le k\le N(T)-J+1}(S_{k+J} - S_k) \ge (1-\varepsilon)\mu b_T\right\}$$
$$\supset \left\{\max_{0\le k\le K-J+1}(S_{k+J} - S_k) \ge (1-\varepsilon)\mu b_T), N(T) \ge K\right\} \supset A_T B_T, \quad (5.12)$$

for all sufficiently large T, where

$$A_T = \{S_{K+1} - S_{K-J+1} \ge (1-\varepsilon)\mu b_T\} \quad \text{and} \quad B_T = \{N(T) \ge K\}.$$

Let $\{T_k\}$ be a sequence of positive numbers such that $T_k \nearrow \infty$ as $k \to \infty$. By the SLLN for sums and

$$B_T = \{R_K \le T\} = \left\{S_K \ge -\frac{K\delta\mu}{1-\delta}\right\},$$

w.p. 1 events B_{T_k} occur for all sufficiently large k. Hence, we only need to check that $P(A_{T_k}\ i.o.) = 1$.

Define d_T by formula (5.10). Making use of (5.3), we have $\mu b_T = c_T\gamma(d_T/c_T) \le a\mu c_T$ for all sufficiently large T, where $a < 1$. Hence,

$$J = [c_T - \varepsilon b_T] \ge [(1-\varepsilon a)c_T]$$

and

$$(1-\varepsilon)c_T \le (J+1)(1-\varepsilon)(1-\varepsilon a)^{-1} \le J(1-\varepsilon_1)^2$$

for all sufficiently large T, where $\varepsilon_1 > 0$.

Using (2.70) and the concavity of $\gamma(x)$, we get

$$\begin{aligned} P(A_T) &\ge P(S_J \ge (1-\varepsilon)\mu b_T) \ge P\left(S_J \ge (1-\varepsilon_1)^2 J\gamma\left(\frac{d_T}{c_T}\right)\right) \\ &\ge P\left(S_J \ge (1-\varepsilon_1)J\gamma\left((1-\varepsilon_1)\frac{d_T}{c_T}\right)\right) \ge \mathrm{e}^{-(1-\varepsilon_2)d_T} \end{aligned} \tag{5.13}$$

for all sufficiently large T.

Put $T_1 = 1$,

$$T_{k+1} = \min\{T : T > T_k, K(T_{k+1}) - J(T_{k+1}) = K(T_k)\}$$

for $k \in \mathbf{N}$. In the same way as in the proof of Theorem 1.2, we prove that the series $\sum_k P(A(T_k))$ diverges. Since the events $\{A(T_k)\}$ are independent, the Borel–Cantelli lemma implies $P(A(T_k)\ \text{i.o.}) = 1$.

Assume now that $a_T/T \to 1$. Then $d_T/c_T \to 0$ and (5.3) yields that $b_T = o(c_T)$. It follows that $c_T \sim a_T/\mu \sim T/\mu$.

By (5.7), we have

$$\{U_T \ge (1-2\varepsilon)b_T\} \supset \{S_J \ge (1-\varepsilon)\mu b_T\}$$

for all sufficiently large T.

Take $\theta > 1$. Put $T_k = \theta^k$ and $J_k = J(T_k)$ for $k \in \mathbf{N}$. Denote

$$C_k = \{S_{J_k} - S_{J_{k-1}} \ge (1-0.5\varepsilon)\mu b_{T_k}\}, \quad D_k = \{S_{J_{k-1}} \ge -0.5\varepsilon\mu b_{T_k}\}.$$

The relation $J_{k-1} \sim c_{T_{k-1}}$ as $k \to \infty$ and condition (5.4) yield that $P(D_k) \ge q$ for all sufficiently large k.

The following pairs of events are independent: C_k and D_k, C_k and $D_k\overline{C_{k-1}D_{k-1}}$, C_k and $D_k\overline{C_{k-1}D_{k-1}C_{k-2}D_{k-2}}, \ldots$

Since $c_{T_k} \sim (J_k - J_{k-1})\theta/(\theta-1)$ as $k \to \infty$, we have

$$P(C_k) \geq P\left(S_{J_k - J_{k-1}} > (1+\tau)(1-0.5\varepsilon)\frac{\theta}{\theta-1}(J_k - J_{k-1})\gamma\left(\frac{d_{T_k}}{c_{T_k}}\right)\right)$$

for all sufficiently large k. Choose large θ and small τ such that $(1+\tau)(1-0.5\varepsilon)\theta/(\theta-1) = (1-\rho)^2$, where $\rho > 0$. Using the concavity and the definition of $\gamma(x)$ and relation (2.70) in the same way as in the proof of (5.9), we obtain

$$P(C_k) \geq \exp\left\{-(1-\rho)(1+\eta)(J_k - J_{k-1})\frac{d_{T_k}}{c_{T_k}}\right\} \geq k^{-(1-\rho)(1+\eta)(1+\xi)}$$

for all sufficiently large k. Choosing of small enough η and ξ implies the divergence of the series $\sum\limits_k P(C_k)$.

By Lemma 1.3, we have

$$P(C_k D_k \ \text{ i.o.}) \geq q > 0.$$

It follows that

$$P(S_{J_k} \geq (1-\varepsilon)\mu b_{T_k} \ \text{ a.s. }) > 0.$$

By Kolmogorov's 0 or 1 law, the last probability equals to 1. This yields that

$$P(U_{T_k} \geq (1-2\varepsilon)b_{T_k} \ \text{ a.s.}) = 1.$$

Taking into account that

$$\{u_T \geq (1-\varepsilon)c_T\} \supset \left\{U_T \geq (1-\varepsilon)\left(c_T - \frac{a_T}{\mu}\right)\right\},$$

we get the result for u_T. □

Turn to lower bounds for sufficiently slowly increasing functions c_T.

Theorem 5.3. *If the conditions of Theorem 5.2 hold and* $\log\log T = o(\log(T/c_T))$, *then*

$$\liminf \frac{U_T}{b_T} \geq 1 \qquad a.s.$$

One can replace U_T *by* u_T *in the last relation.*

Proof. Under the notations of the proof of Theorem 5.2, write (5.12) in

$$\{U_T < (1-2\varepsilon)b_T\} \subset E_T \cup F_T$$

for all sufficiently large T, where $E_T = \overline{A_T}$ and $F_T = \overline{B_T}$.

By the SLLN for sums and

$$F_T = \{R_K > T\} = \left\{S_K < -\frac{K\delta\mu}{1-\delta}\right\},$$

w.p. 1 the events F_T do not occur for all sufficiently large T. Hence, it is sufficient to prove the same for the events E_T.

We have

$$P(E_T) \le P\left(\bigcap_{m=0}^{[K/J]} (S_{(m+1)J} - S_{mJ}) < (1-\varepsilon)\mu b_T\right)$$
$$\le (1 - P(S_J \ge (1-\varepsilon)\mu b_T))^{[K/J]} \le \mathrm{e}^{-[K/J]P(S_J \ge (1-\varepsilon)\mu b_T)}. \quad (5.14)$$

Applying (5.13), the definitions of K and J and the condition $\log\log T = o(\log T/c_T)$, we get

$$P(E_T) \le T^{-3} \quad (5.15)$$

for all sufficiently large T.

Put $T_n = \mu n/(1-\delta)$ for $n \in \mathbf{N}$. For every n with $J(T_n) < J(T_{n+1})$, put

$$T_n(j) = \inf\{T \in [T_n, T_{n+1}) : J(T) = j\}, \quad j \in I_n = (J(T_n), J(T_{n+1})].$$

For n with $J(T_n) = J(T_{n+1})$, put

$$T_n(j) = T_{n+1}, \quad j \in I_n = \{J(T_{n+1})\}.$$

By the definition of $T_n(j)$, we have $E_T \subset E_{T_n(j+1)}$ for $T \in (T_n(j), T_n(j+1)]$. Further,

$$\sum_{n=N}^{\infty} \sum_{j \in I_n} P(E_{T_n(j+1)}) \le \sum_{n=N}^{\infty} \sum_{j \in I_n} (T_n(j+1))^{-3}$$
$$\le \sum_{n=N}^{\infty} (J(T_{n+1}) - J(T_n) + 1)T_n^{-3}.$$

Here N is chosen large enough to satisfy (5.15). In the same way as in the proof of Theorem 5.2, we get $c_T \ge J(T) \ge [(1-a\varepsilon)c_T]$. Hence,

$$J(T_{n+1}) - J(T_n) \le c_{T_{n+1}} - (1-a\varepsilon)c_{T_n} + 1.$$

Since $c_{T_n}/c_{T_{n+1}} \le T_{n+1}/T_n \to 1$ and $c_{T_n} = O(T_n)$ as $n \to \infty$, we have

$$J(T_{n+1}) - J(T_n) = O(T_n)$$

as $n \to \infty$. Then the series $\sum_{n=N}^{\infty} \sum_{j \in I_n} P(E_{T_n(j+1)})$ converges. By the Borel–Cantelli lemma, w.p. 1 a finite number of events $E_{T_n(j+1)}$ occurs. It follows that w.p. 1 the events E_T do not occur for all sufficiently large T.

Making use of

$$\{u_T \ge (1-\varepsilon)c_T\} \supset \left\{U_T \ge (1-\varepsilon)\left(c_T - \frac{a_T}{\mu}\right)\right\},$$

we obtain the result for u_T. □

Theorems 5.1–5.3 and the inequality $U_n \le W_n$ imply the next result.

Theorem 5.4. *If the conditions of Theorems 5.1 and 5.2 hold, then*

$$\limsup \frac{W_T}{b_T} = \limsup \frac{U_T}{b_T} = \limsup \frac{u_T}{b_T} = 1 \quad a.s. \tag{5.16}$$

If the conditions of Theorems 5.1 and 5.3 hold, then

$$\lim \frac{W_T}{b_T} = \lim \frac{U_T}{b_T} = \lim \frac{u_T}{b_T} = 1 \quad a.s. \tag{5.17}$$

5.2 Corollaries of the Universal Strong Laws

We state a numbers of important corollaries of Theorem 5.4 in this section.

We start with the case $a_T = O(\log T)$. It is not difficult to check that relation (5.1) is equivalent to

$$c_T = a_T \beta\left(\frac{\log(T/c_T) + \log\log T}{a_T}\right), \tag{5.18}$$

where $\beta(x)$ is the inverse function to $\kappa(u) = u\zeta(\mu - u^{-1})$, $u \ge \mu^{-1}$. It follows from the properties of $\zeta(x)$ that $\beta(0) = 1/\mu$, $\beta(x) \nearrow \infty$ as $x \to \infty$ and $\beta(x)$ is concave.

The first result is as follows.

Theorem 5.5. *If* $a_T = c\log T$, $c > c_0$, *then*

$$\lim \frac{W_T}{(\beta(1/c) - 1/\mu)a_T} = \lim \frac{U_T}{(\beta(1/c) - 1/\mu)a_T} = \lim \frac{u_T}{\beta(1/c)a_T} = 1 \quad a.s.$$

Theorem 5.5 is the Erdős–Rényi law for renewal processes.

Remember that the norming functions are calculated from different parts of formula (5.2) for U_T and u_T. Simple calculations show that the relations with U_T and u_T in Theorem 5.5 are equivalent.

Proof. Consider U_T. Define d_T by formula (5.10). By the conditions of theorem, $\gamma(d_T/c_T)$ is separated from zero and μ for all sufficiently large T. Hence,

$$c_T = \frac{a_T}{\mu - \gamma(d_T/c_T)} = O(\log T).$$

Making use of (5.18) and the concavity of $\beta(x)$, we get

$$c_T = a_T\beta\left((1+o(1))\frac{\log T}{a_T}\right) \sim a_T\beta\left(\frac{1}{c}\right).$$

It follows that $b_T \sim a_T(\beta(1/c) - 1/\mu)$.

The case of u_T follows from the above provided $c_T = b_T$. □

For $a_T = o(\log T)$, we have the next result.

Theorem 5.6. *Assume that $a_T/\log T \searrow 0$ and one of the following conditions holds:*

1) $P(X = \mu) > 0$;

2) $P(X = \mu) = 0$ and $\mu - \gamma(-\log P(X > \mu - x)) \sim x$ as $x \searrow 0$.

Then relation (5.17) holds with $b_T = a_T\beta(\log T/a_T)$.

Our formula for norming sequences works for $a_T = o(\log T)$ as well. To check this, we only need to prove that

$$b_T \sim a_T\beta\Big(\frac{\log T}{a_T}\Big). \tag{5.19}$$

Define d_T by formula (5.10). Check that $\gamma(d_T/c_T) \to \mu$. Assume that there exists a sequence $\{T_k\}$, $T_k \nearrow \infty$ as $k \to \infty$, such that $\mu - \gamma(d_{T_k}/c_{T_k}) > \rho > 0$ for some ρ. Then $d_{T_k}/c_{T_k} < a < \infty$ for some a. Hence, $\log T_k/c_{T_k} < 2a$ for all sufficiently large k. It follows that

$$a_{T_k} = \left(\mu - \gamma\left(\frac{d_{T_k}}{c_{T_k}}\right)\right) c_{T_k} > \frac{\rho}{2a}\log T_k$$

for all sufficiently large k, which contradicts to $a_T = o(\log T)$.

Relation (5.1) implies that $a_T = o(c_T)$ and $c_T \sim b_T$. In particular, the latter means that in Theorem 5.6, the norming functions for U_T and u_T coincide since the centering a_T/μ is negligible.

If $P(X = 0) = 0$, then $\gamma(d_T/c_T) \to \mu$ implies that $d_T/c_T \to \infty$. Hence, $c_T = o(\log T)$ and $b_T = o(\log T)$. By (5.18) and concavity of $\beta(x)$, we have

$$c_T = a_T\beta\left((1+o(1))\frac{\log T}{a_T}\right) \sim a_T\beta\left(\frac{\log T}{a_T}\right).$$

This gives (5.19).

If $P(X = 0) > 0$, then $\gamma(d_T/c_T) \to \mu$ yields $d_T/c_T \to 1/c_0$. Then $c_T \sim c_0\log T$ and $b_T \sim c_0\log T$. But the left-hand side of (5.19) is equivalent to $c_0\log T$ as well.

Proof. Suppose first that condition 1) holds. Put $b_T = c_0 \log T$. Using the notations from the proof of Theorem 5.3, we get inequality (5.14) and $J \sim (1-\varepsilon)b_T$. Taking into account that $c_0 = -\log P(X=\mu)$, we have

$$P(S_J \geq (1-\varepsilon)\mu b_T) \geq (P(X=\mu))^J = \mathrm{e}^{-J/c_0} \geq T^{-1+\varepsilon/2}$$

for all sufficiently large T. It follows that

$$P(E_T) \leq \mathrm{e}^{-T^{\varepsilon/4}}$$

for all sufficiently large T. The remainder coincides with that of the proof of Theorem 5.3.

Suppose now that condition 2) holds. Define b_T by $b_T = a_T\beta(\log T/a_T)$. Take $\varepsilon, \delta \in (0,1)$. Put $K = K(T) = [(1-\delta)T/\mu]$, $J = J(T) = [(1-\varepsilon)b_T]$ and $T_n = \mu n/(1-\delta)$ for $n \in \mathbf{N}$. By (5.12), we have

$$\begin{aligned}
&P(u_T < (1-2\varepsilon)b_T) \leq P(u_T < (1-\varepsilon)b_T - 2)\\
&\leq P\left(\min_{0\leq k\leq N(T)-J}(R_{k+J} - R_k) \geq a_T\right)\\
&\leq P\left(\min_{0\leq k\leq N(T)-J}(R_{k+J} - R_k) \geq a_{T_{n-1}}\right)
\end{aligned}$$

for $T_{n-1} \leq T \leq T_n$. Put $J_n = J(T_n)$ and

$$p_n = P\left(\min_{0\leq k\leq n-J_n}(R_{k+J_n} - R_k) \geq a_{T_{n-1}}\right)$$

for $n \in \mathbf{N}$. Note that $K(T_n) = n$ for all n. In the same way as in the proof of Theorem 5.3, the SSLN for sums yields that the result follows from $\sum_n p_n < \infty$. Check the latter.

Since $a_{T_n}/a_{T_{n-1}} \leq \log T_n/\log T_{n-1} \to 1$ as $n \to \infty$, we have

$$\begin{aligned}
p_n &\leq P\left(\min_{0\leq k\leq n-J_n}(R_{k+J_n} - R_k) \geq (1-\varepsilon)^{3/2}a_{T_n}\right)\\
&\leq (P(R_J \geq (1-\varepsilon)^{3/2}a_{T_n}))^{[n/J_n]} \leq \mathrm{e}^{-[n/J_n]P(R_{J_n}\geq(1-\varepsilon)^{3/2}a_{T_n})}
\end{aligned}$$

for all sufficiently large n. Further,

$$\begin{aligned}
&P(R_{J_n} \geq (1-\varepsilon)^{3/2}a_{T_n}) \geq \left(P\left(Y \geq (1-\varepsilon)^{3/2}\frac{a_{T_n}}{J_n}\right)\right)^{J_n}\\
&\geq \mathrm{e}^{-J_n \log P(X>\mu-(1-\varepsilon)^{3/2}a_{T_n}/J_n)} \geq T_n^{-1+\varepsilon}
\end{aligned}$$

for all sufficiently large n. This yields $\sum_n p_n < \infty$. □

Turn to the case $a_T/\log T \to \infty$. Then the behaviour of u_T is quite different from that of W_T and U_T. For example, if $EX^2 < \infty$ and $a_T = T$, then the result for U_T turns to the LIL for renewal processes while that for u_T turns to the SLLN. Therefore, we deal with W_T and U_T separately from u_T.

We start with the case $EX^2 < \infty$. Our first result is as follows.

Theorem 5.7. *If $\sigma^2 = EX^2 < \infty$, then*

$$\limsup \frac{W_T}{b_T} = \limsup \frac{U_T}{b_T} = 1 \quad a.s., \tag{5.20}$$

where

$$b_T = \frac{\sigma}{\mu^{3/2}} \sqrt{2a_T \left(\log \frac{T}{a_T} + \log\log T\right)}.$$

If additionally $\log\log T = o(\log(T/a_T))$, then

$$\lim \frac{W_T}{b_T} = \lim \frac{U_T}{b_T} = 1 \quad a.s. \tag{5.21}$$

Note that for Y with the exponential distribution, we have $\sigma^2 = DY = \mu^2$. Hence, for the Poisson process, the constant in the above formula for b_T is $\mu^{-1/2}$.

Theorem 5.7 is the Csörgő–Révész laws for renewal processes. It yields the LIL for increments of renewal processes.

Corollary 5.1. *If the conditions of Theorem 5.7 hold and $\log(T/a_T)/\log\log T \to c \geq 0$, then*

$$\limsup \frac{W_T}{\sqrt{2a_T \log\log T}} = \limsup \frac{U_T}{\sqrt{2a_T \log\log T}} = \sqrt{1+c}\,\frac{\sigma}{\mu^{3/2}} \quad a.s.$$

For $a_T = T$, the last relation is the LIL for $N(T)$. The condition $EX^2 < \infty$ is necessary in this case.

Note that Theorem 5.7 may be derived by an application of the strong approximation of renewal processes by the Wiener process and the results for increments of the Wiener processes. Nevertheless, this method can not be used for small increments and for the case $EX^2 = \infty$.

Assume now that $EX^2 = \infty$. For a domain of attraction of the normal law, the result is as follows.

Theorem 5.8. *If $F \in D(2)$, then (5.20) holds with*

$$b_T = a_T \hat{m}\left(\hat{f}^{-1}\left(\frac{\log(T/a_T) + \log\log T}{a_T}\right)\right), \tag{5.22}$$

where

$$\hat{m}(h) = hG\left(\frac{1}{h}\right), \quad \hat{f}(h) = \frac{h^2}{2}G\left(\frac{1}{h}\right), \quad G(x) = \int_{-x}^{0} u^2 dF(u), \;\; x > 0.$$

($G(x) \in SV_\infty$.) If additionally $\log\log T = o(\log(T/a_T))$, *then (5.21) holds with* b_T *from (5.22).*

If $\log(T/a_T)/\log\log T \to c \geq 0$ in Theorem 5.8, then we have an analogue of Corollary 5.1 with an additional slowly varying multiplier in the norming function. For $a_T = T$, the LIL follows as well.

We now deal with X from domains of attraction of completely asymmetric stable laws with $\alpha \in (1,2)$. For domains of normal attraction, we have the following result.

Theorem 5.9. *If* $F \in DN(\alpha)$, $\alpha \in (1,2)$, *then relation (5.20) holds with*

$$b_T = \mu^{-1-1/\alpha}\lambda^{-\lambda}a_T^{1/\alpha}\left(\log\frac{T}{a_T} + \log\log T\right)^{\lambda}, \tag{5.23}$$

where $\lambda = (\alpha-1)/\alpha$. *If additionally* $\log\log T = o(\log(T/a_T))$, *then (5.21) holds with* b_T *from (5.23).*

Turn to the LIL for increments of renewal processes in this case. We have the following result.

Corollary 5.2. *If the conditions of Theorem 5.9 hold and* $\log(T/a_T)/\log\log T \to c \geq 0$, *then*

$$\limsup \frac{W_T}{\lambda^{-\lambda}a_T^{1/\alpha}(\log\log T)^{\lambda}} =$$
$$\limsup \frac{U_T}{\lambda^{-\lambda}a_T^{1/\alpha}(\log\log T)^{\lambda}} = (1+c)^{\lambda}\,\mu^{1+1/\alpha} \qquad a.s.$$

For $a_T = T$, Corollary 5.2 yields the LIL for renewal processes in this case.

Turning to domains of non-normal attraction, we get the next result.

Theorem 5.10. *If* $F \in D(\alpha)$, $\alpha \in (1,2)$, *then relation (5.20) holds with*

$$b_T = a_T\hat{m}\left(\hat{f}^{-1}\left(\frac{\log(T/a_T) + \log\log T}{a_T}\right)\right), \tag{5.24}$$

where

$$\hat{m}(h) = \frac{\alpha\Gamma(2-\alpha)}{\alpha-1}h^{\alpha-1}G\left(\frac{1}{h}\right), \quad \hat{f}(h) = \Gamma(2-\alpha)h^{\alpha}G\left(\frac{1}{h}\right),$$

$G(x) = x^{\alpha}F(-x)$, $x > 0$. *($G(x)SV_\infty$.) If additionally* $\log\log T = o(\log(T/a_T))$, *then (5.21) holds with* b_T *from (5.24).*

Proof. Since $a_T/\log T \to \infty$, we have $d_T/c_T \to 0$. By (5.1), we get $c_T \sim a_T/\mu$ and $b_T = o(a_T)$. Theorems 5.7–5.10 now follow from (5.3) and the results on asymptotics of $\gamma(x)$ at zero from Chapter 2. □

If $\log(T/a_T)/\log\log T \to c \geq 0$ in Theorem 5.10, then we obtain an analogue of Corollary 5.2. For this case, a slowly varying multiplier appears in the norming function. For $a_T = T$, this yields the LIL.

We end this section with a result for u_T that includes the SLLN for renewal processes.

Theorem 5.11. *If* $a_T/\log T \to \infty$, *then*

$$\limsup \frac{u_T}{a_T} = \frac{1}{\mu} \qquad a.s.$$

If $\log\log T = o(\log(T/a_T))$ *or* $a_T = T$, *then one can replace* lim sup *by* lim *in the last relation.*

Proof. The condition $a_T/\log T \to \infty$ yields that $d_T/c_T \to 0$. By (5.1), we have $b_T = c_T \sim a_T/\mu$. By Theorem 5.2, the result follows besides the case $a_T = T$. For $a_T = T$, we have $u_T = N(T)$. Then we get the result from the duality $\{N(T) < k\} = \{R_k > T\}$, $k \in \mathbf{N}$, and the SLLN for sums R_n. □

5.3 Bibliographical Notes

Limit theorems for renewal processes may be found in [Feller (1971)] and [Gut (2009)].

Limit theorems for increments of renewal processes are derived in [Steinebach (1982, 1986, 1991)], [Deheuvels and Steinebach (1989)], [Bacro *et al.* (1987)], [Frolov (2003f,g)]. The universal strong laws for Compound renewal processes are obtained by [Frolov (2007)].

Chapter 6

Increments of Sums of Independent Random Variables over Head Runs and Monotone Blocks

Abstract. We derive universal strong laws for increments of sums of independent random variables over head runs and monotone blocks. They yield the SLLN, the LIL, the Erdős–Rényi law and the Csörgő–Révész laws.

6.1 Head Runs and Monotone Blocks

We start with notions of head runs and monotone blocks.

Let $\{Y_n\}$ be a sequence of independent Bernoulli random variables. A series of ones is called a head run. A.s. asymptotic behaviour of the longest head run in $Y_1, Y_2, \dots, Y_n$ is well known. In the sequel, we permanently use this asymptotic and, therefore, we need the following result.

Lemma 6.1. *Put* $p = P(Y_1 = 1)$ *and*

$$L_n^{hr} = \max\{k : Y_{i+1} = \cdots = Y_{i+k} = 1 \text{ for some } i,\ 0 \le i \le n-k\}.$$

If $p \in (0,1)$, *then*

$$\lim \frac{L_n^{hr}}{\log n/log(1/p)} = 1 \quad a.s.$$

In what follows, we assume that $n \to \infty$ in $\sim$, $\to$, lim sup, lim inf, o, O provided it is not pointed otherwise.

Proof. Put $l_n = \log n/\log(1/p)$. For $\varepsilon > 0$, we have

$$\begin{aligned} P_n = P(L_n^{hr} \ge (1+\varepsilon)l_n) \le nP(Y_1 = \cdots = Y_{[(1+\varepsilon)l_n]} = 1) \qquad (6.1) \\ \le np^{[(1+\varepsilon)l_n]} \le n^{-\varepsilon/2} \end{aligned}$$

for all sufficiently large n. In the last inequality, we have used the definition of l_n. Put $n_k = 2^k$ for all $k \in \mathbf{N}$. Then the series $\sum_k P_{n_k}$ converges. By the Borel–Cantelly lemma, we have

$$\limsup_{k\to\infty} \frac{L_{n_k}^{hr}}{l_{n_k}} \le 1+\varepsilon \quad \text{a.s.}$$

For $n_k \le n \le n_{k+1}$, we have

$$\frac{L_n^{hr}}{l_n} \le \frac{L_{n_{k+1}}^{hr}}{l_{n_{k+1}}} \frac{l_{n_{k+1}}}{l_{n_k}} \le \frac{L_{n_{k+1}}^{hr}}{l_{n_{k+1}}}(1+\varepsilon)$$

for all sufficiently large k. Hence,

$$\limsup \frac{L_n^{hr}}{l_n} \le (1+\varepsilon)^2 \quad \text{a.s.}$$

Passing to the limit as $\varepsilon \to 0$ in this relation, we get the upper bound.

Turn to the lower bound. For $\varepsilon > 0$, we have

$$Q_n = P(L_n^{hr} \le (1-\varepsilon)l_n) \le \left(1 - P(Y_1 = \cdots = Y_{[(1-\varepsilon)l_n]} = 1)\right)^{[n/l_n]-1} \tag{6.2}$$
$$\le \exp\left\{-\left[\frac{n}{l_n} - 1\right] p^{[(1-\varepsilon)l_n]}\right\} \le \exp\left\{-n^{\varepsilon/2}\right\}$$

for all sufficiently large n. In the last inequality, we have used the definition of l_n. It follows that the series $\sum_n Q_n$ converges. By the Borel–Cantelli lemma, we get

$$\liminf \frac{L_n^{hr}}{l_n} \ge 1-\varepsilon \quad \text{a.s.}$$

Passing to the limit as $\varepsilon \to 0$ in this relation, we get the lower bound. □

Turn to monotone blocks.

Let $\{Y_n\}$ be a sequence of independent, continuous random variables. A series $Y_{i+1} < Y_{i+2} < \cdots < Y_{i+k}$ with $0 \le i \le i+k \le n$ is called a monotone block (or an increasing run) of length k in $Y_1, Y_2, \ldots, Y_n$. The asymptotic of the length of the longest monotone block in $Y_1, Y_2, \ldots, Y_n$ will also be used in the sequel. Hence, we need the next result.

Lemma 6.2. *Put*

$$L_n^{mb} = \max\{k : Y_{i+1} < \cdots < Y_{i+k} \text{ for some } i,\ 0 \le i \le n-k\}.$$

Then

$$\lim \frac{L_n^{mb}}{\log n/\log\log n} = 1 \quad a.s.$$

Proof. Note that $P(Y_1 < Y_2 < \cdots < Y_k) = 1/k!$ for all k.

Put $l_n = \log n / \log\log n$. For $\varepsilon > 0$, we have the following analogue of (6.1):

$$P(L_n^{mb} \geq (1+\varepsilon)l_n) \leq nP(Y_1 < \cdots < Y_{[(1+\varepsilon)l_n]}).$$

This yields

$$P(L_n^{mb} \geq (1+\varepsilon)l_n) \leq \frac{n}{[(1+\varepsilon)l_n]!} \leq n^{-\varepsilon/2} \tag{6.3}$$

for all sufficiently large n. The last inequality follows from the Stirling formula

$$n! \sim \sqrt{2\pi n}\, n^n \,\mathrm{e}^{-n}$$

and the definition of l_n. We get instead of (6.2) that

$$P(L_n^{mb} \leq (1-\varepsilon)l_n) \leq \left(1 - P(Y_1 < \cdots < Y_{[(1-\varepsilon)l_n]})\right)^{[n/l_n]-1}.$$

It implies that

$$P(L_n^{mb} \leq (1-\varepsilon)l_n) \leq \exp\left\{-\left[\frac{n}{l_n} - 1\right]\frac{1}{[(1-\varepsilon)l_n]!}\right\} \leq \exp\left\{-n^{\varepsilon/2}\right\} \tag{6.4}$$

for all sufficiently large n. We have used the definition of l_n and the Stirling formula in the last inequality. Inequalities (6.3) and (6.4) imply the result in the same way as in the proof of Lemma 6.1. □

In literature, one can find results for another interesting cases as well. Rejecting the assumption of continuity for Y, we arrive at a quite different problem with the lengths of the longest monotone blocks. Moreover, we can also investigated non-strictly monotone blocks which are defined with a replacement of signs $<$ by $\leq$. Further generalizations are blocks for d-dimensional Y's or random fields. (See, [Frolov and Martikainen (1999, 2001)] and references therein). Below we mention another generalizations of monotone blocks as well.

6.2 Increments of Sums over Head Runs and Monotone Blocks

We are going to investigate a.s. limit behaviour of increments of sums of i.i.d. random variables over head runs and monotone blocks in an accompanying sequence of i.i.d. random variables. To this end, we deal with the settings as follows.

Let $(X, Y), (X_1, Y_1), (X_2, Y_2), \ldots$ be a sequence of i.i.d. random vectors. Note that X and Y can be dependent and the case $X = Y$ can be considered as well. Put

$$S_n = X_1 + X_2 + \cdots + X_n, \quad S_0 = 0.$$

We will investigate the a.s. asymptotic behaviour for maxima

$$M_n = \max_{0 \leq k \leq n - j_n} (S_{k+[j_n]} - S_k) I_{\{u \leq Y_{k+1} \leq \cdots \leq Y_{k+[j_n]} \leq v\}},$$

where u, v are fixed, $-\infty \leq u \leq v \leq +\infty$ and $\{j_n\}$ is such that $1 \leq j_n \leq n$ for all n. Here I_B is the indicator of the event B and $[x]$ is the integer part of the number x. Note that j_n may be a random variable.

Our notation for the lengths of increments is different from that in the previous chapters. This is to emphasize that we will never have j_n so large as a_n. Formally, we assume that $j_n \leq n$, but j_n will always have a logarithmic or smaller order. Indeed, all indicators in M_n are zeros when j_n is greater than the length of the longest block of Y's. By Lemmas 6.1 and 6.2, these lengths have a logarithmic order of growth w.p. 1 in the cases we concern with.

Our aim is to find a sequence of positive numbers $\{b_n\}$ such that either

$$\limsup_{n \to \infty} \frac{M_n}{b_n} = 1 \quad \text{a.s.},$$

or the latter holds with lim sup replaced by lim when it is possible.

The maximum M_n has an interesting interpretation in game settings. Assume that X_n is a gain (the negative gain is a loss) of a player in n-th repetition of a game of chance and Y_n is a characteristic of a success in n-th repetition. For example, if $Y_n = I_{\{X_n > 0\}}$, then n-th repetition is successful provided the gain is positive. Note that sometimes we can only define a success for a successive series of repetitions. Indeed, try to tell that n-th repetition is successful if the gain in this repetition is greater than the gain in the previous one. Then $Y_n = I_{\{X_n > X_{n-1}\}}$ and the vectors (X_n, Y_n) are not independent. This difficulty can be eliminated as follows. Put $Y_n = X_n$ and say that a series of repetitions of the game is successful when the gain in every next repetition is greater than that in the previous one. In these settings, M_n is a maximal gain in successful series of lengths j_n.

Investigations of asymptotics for M_n is of interest in probability and actuarial and financial mathematics. In the last case, it allows to estimate possible losses on time subintervals (see, for example [Binswanger and Embrechts (1994)]). Moreover, the procedure of rarefying naturally appears in there (see [Embrechts and Clüppelberg (1993)]). For example, insurance companies pay compensations after investigations and some claims are rejected. In our settings, we put $Y_n = 1$, when n-th claim is accepted and $Y_n = 0$ otherwise. Maxima M_n for $u = v = 1$ is a maximal aggregate amount of compensations in time j_n over series of accepted claims.

The above settings are very general. Interesting partial appear under various additional assumptions on distributions of the vector (X, Y), the sequence $\{j_n\}$, u and v.

Assume first that $u = v = 1$, $P(Y = 1) = 1 - P(Y = 0) = p \in (0, 1]$. If X is a non-degenerate random variables and $p = 1$, then M_n coincides with U_n which has been investigated in Chapter 3. If $X = 1$ a.s., $p \in (0, 1)$ and j_n is the length of the longest head run in $Y_1, Y_2, \ldots, Y_n$. Then $M_n = j_n = L_n^{hr}$ which behaviour is described by Lemma 6.1. If X is a non-degenerate random variables and $p < 1$, then M_n is the maximal increments of sums S_n over head runs. This case will be considered below.

Assume further that $u = -\infty$, $v = +\infty$ and Y has a continuous distribution. If $X = 1$ a.s. and j_n is the length of the longest monotone block in $Y_1, Y_2, \ldots, Y_n$, then $M_n = j_n = L_n^{mb}$ which the asymptotic is found in Lemma 6.2 for. If X is a non-degenerate random variables, then M_n is the maximal increments of sums S_n over monotone blocks and its limiting behaviour will be described below.

Assume finally that $-\infty < u < v < +\infty$ and Y has a continuous distribution. Then M_n is the maximal increments of sums S_n over monotone blocks in the interval (u, v). This case will be discussed below as well.

In the next section, we derive the universal laws for M_n when X is a non-degenerate random variable and Y is either continuous, or $P(Y = 1) = 1 - P(Y = 0) = p \in (0, 1)$. Further, we concern with various corollaries of the universal laws.

6.3 The Universal Strong Laws

Dealing with M_n, we only consider sequences $\{j_n\}$ with $j_n \le L_n$ a.s., where L_n is the length of the longest sequence of Y's taking its value in $[u, v]$, i.e.

$$L_n = \max\{k : u \le Y_{i+1} \le \cdots \le Y_{i+k} \le v \text{ for some } i, \ 0 \le i \le n - k\}.$$

Consider two following cases.

A) Assume that $u = v = 1$ and $P(Y = 1) = p = 1 - P(Y = 0) \in (0, 1)$. Then $L_n = L_n^{hr}$ is the length of the longest head run in $Y_1, Y_2, \ldots, Y_n$ and $L_n \sim \log n / \log(1/p)$ a.s. by Lemma 6.1.

B) Suppose that $-\infty \le u < v \le +\infty$ and $P(Y = y) = 0$ for all y. If $u = -\infty$ and $v = +\infty$, then $L_n = L_n^{mb}$ is the length of the longest monotone block in $Y_1, Y_2, \ldots, Y_n$ and $L_n \sim \log n / \log\log n$ a.s. by Lemma 6.2. For finite u and v, the asymptotic of L_n is the same provided $p = P(u < Y < v) > 0$. This is clear from the proof of Lemma 6.2 and the equality $P(u < Y_1 < Y_2 < \cdots < Y_k < v) = p^k/k!$ for all k. One can check that the additional multiplier in the last formula is negligible in analogues of bounds (6.3) and (6.4) for this case.

The behaviour of L_n has been investigated for other blocks and distribution of Y's as well (see [Frolov and Martikainen (1999)]). But large deviations asymptotics are known for head runs and monotone blocks (see Lemmas 6.3–6.5 below). Hence, we restrict our attention by cases A) and B).

For $n \in \mathbf{N}$, put

$$l_n = \frac{\log n}{\log(1/p)}$$

in case A) and let l_n be a solution of the equation

$$\sqrt{2\pi l}\, l^l \,\mathrm{e}^{-l} = n$$

in case B). Note that the left-hand part of this equation is the asymptotic of $l!$ from Stirling's formula. It is not difficult to check that in this case,

$$l_n \sim \frac{\log n}{\log\log n}.$$

Put $p = P(u \le Y \le v) > 0$. Let $\bar{X}$ be a random variable with the d.f. $F(x) = P(X < x | u \le Y \le v)$. Note that for $u = -\infty$ and $v = +\infty$, the conditional d.f. $F(x)$ coincides with the d.f. of X.

In the sequel, we will assume that the random variable $\bar{X}$ is non-degenerate, $E\bar{X} \ge 0$ and

$$h_0 = \sup\{h : \varphi(h) = Ee^{h\bar{X}} < \infty\} > 0.$$

Put $\omega = \operatorname{esssup}\bar{X}$. For $0 < h < h_0$, put

$$m(h) = \frac{\varphi'(h)}{\varphi(h)}, \quad \sigma^2(h) = m'(h), \quad f(h) = hm(h) - \log\varphi(h).$$

Denote

$$\zeta(z) = \sup_{h \ge 0, \varphi(h) < \infty} \{zh - \log\varphi(h)\}, \quad \gamma(x) = \sup\{z : \zeta(z) \le x\}.$$

Properties of the functions $m(h)$, $\sigma^2(h)$, $\zeta(z)$ and $\gamma(x)$ have been discussed in Chapter 2. We remember them taking into account that these function are constructed from a conditional distribution. We have

$$m(0) = E\bar{X}, \ \ m(h) \nearrow A \ \text{as} \ h \nearrow h_0, \ \ f(0) = 0, \ \ f(h) \nearrow \frac{1}{c_0} \ \text{as} \ h \nearrow h_0,$$

$$\sigma^2(h) > 0, \quad \sigma^2(0) = D\bar{X} \ \text{if} \ D\bar{X} < \infty,$$

$$\zeta(E\bar{X}) = 0, \ \ \zeta(z) \nearrow, \ \ \zeta(z) \ \text{is convex}, \ \ \zeta(z) = +\infty \ \text{for} \ z > A \ \text{if} \ A < \infty,$$

$$\gamma(0) = E\bar{X}, \ \ \gamma(x) \nearrow \ \text{for} \ x < \frac{1}{c_0}, \quad \gamma(x) \ \text{is concave},$$

$$\gamma(x) = \omega \ \text{for} \ x > \frac{1}{c_0} \ \text{if} \ c_0 > 0 \ \text{and} \ \omega < \infty, \ \ \gamma(x) \to \omega \ \text{as} \ x \to \infty.$$

Moreover, we have

$$\zeta(z) = f(m^{-1}(z)) \ \text{for} \ z \in [E\bar{X}, A),$$

$$\gamma(x) = m(f^{-1}(x)) \ \text{for} \ x \in \left[0, \frac{1}{c_0}\right), \tag{6.5}$$

where $m^{-1}(\cdot)$ and $f^{-1}(\cdot)$ are inverse functions to $m(\cdot)$ and $f(\cdot)$ correspondingly.

Let $\{\delta_n\}$ be a sequence of real numbers with $\delta_n \in (0,1)$. In the sequel, we only consider sequences $\{j_n\}$ as follows:

$$j_n = \delta_n l_n.$$

For $n \in \mathbf{N}$, put $i_n = [\delta_n l_n]$,

$$\tau_n = (l_n - i_n) \log \frac{1}{p} \quad \text{in case A)},$$

$$\tau_n = \left(l_n + \frac{1}{2}\right) \log l_n - l_n - \left(i_n + \frac{1}{2}\right) \log i_n + i_n - i_n \log \frac{1}{p} \quad \text{in case B)}.$$

The formula for τ_n is rather complicated for the case B). To realise the asymptotic behaviour of τ_n, we write

$$\tau_n = (l_n - i_n) \log l_n + \left(i_n + \frac{1}{2}\right) (\log l_n - \log i_n) + (i_n - l_n) - i_n \log \frac{1}{p}.$$

For $i_n \to \infty$, the main part of the asymptotic of τ_n is $(1-\delta_n) l_n \log l_n$ for $\{\delta_n\}$ separated from 1 and $(1-\delta_n) l_n \log l_n - \delta_n l_n \log(1/p)$ for $\{\delta_n\}$ with $\delta_n \to 1$. Hence, we will assume that $(1-\delta_n)\log l_n > \log(1/p)$. For the case of monotone blocks, the last equality is trivial by $p = 1$.

Our first result is as follows.

Theorem 6.1. *Assume that $\bar{X}$ is non-degenerate, $E\bar{X} \geq 0$, $h_0 > 0$, and $\{i_n\}$ is a non-decreasing sequence with $i_n \geq 1$ for all n. Then*

$$\limsup \frac{M_n}{b_n} \leq 1 \quad a.s., \tag{6.6}$$

where $b_n = i_n \gamma((\tau_n + \log l_n)/i_n)$.

To prove Theorem 6.1, we need some auxiliary results.

We start with bounds for probabilities of large deviations. Two first Lemmas are variants of Lemmas 2.2 and 2.24 from Chapter 2 which we rewrite in term of random variables with the (conditional) d.f. $F(x)$.

Lemma 6.3. *Assume that the conditions of Theorem 6.1 hold. Let* $\{\bar{X}_i\}$ *be a sequence of i.i.d. random variables with the same distribution as* $\bar{X}$. *Put* $\bar{S}_n = \bar{X}_1 + \cdots + \bar{X}_n$.

Then

$$P(\bar{S}_n \geq nm(h)) \leq e^{-nf(h)},$$

$$P(\bar{S}_n \geq nm(h) - 2\sqrt{n}\sigma(h)) \geq \frac{3}{4}e^{-nf(h)-2\sqrt{n}h\sigma(h)}$$

for all n *and* $h \in (0, h_0)$.

For lower bounds, we have the following result.

Lemma 6.4. *Let* $\{h_n\}$ *be a sequence of real numbers with* $h_n \in (0, h_0)$. *Assume that the conditions of Lemma 6.3 and one of two following conditions hold:*

1) $nf(h_n) \to \infty$ *and* $h_n\sigma(h_n) = o(\sqrt{n}f(h_n))$.

2) $h_n \leq \tilde{h} < h_0$.

Then

$$P\left(\bar{S}_n \geq (1-\varepsilon)nm(h)\right) \geq \frac{3}{4}e^{-nf(h)(1+\delta)}$$

for all $\varepsilon \in (0, 1)$, $\delta > 0$ *and all sufficiently large* n.

The next lemma describes relationships between bounds for large deviations of sums over monotone blocks and head runs and those for sums of i.i.d. random variables with the conditional d.f. $F(x)$.

Lemma 6.5. *If* $u < v$ *and* $P(Y = y) = 0$ *for all* y, *then for all* $x > 0$, *the following relations hold:*

$$P\left(S_n I_{\{u \leq Y_1 \leq \cdots \leq Y_n \leq v\}} \geq x\right) = \frac{p^n}{n!} P\left(\bar{S}_n \geq x\right), \tag{6.7}$$

$$P(M_n \geq x) \leq \frac{np^{j_n}}{j_n!} P\left(\bar{S}_{j_n} \geq x\right), \tag{6.8}$$

$$P(M_n < x) \leq \exp\left\{-\left[\frac{n}{j_n}\right]\frac{p^{j_n}}{j_n!} P\left(\bar{S}_{j_n} \geq x\right)\right\}. \tag{6.9}$$

If $u = v = 1$ *and* $P(Y = 1) = p > 0$, *then for all* $x > 0$, *relation (6.7) holds with a replacement of* $n!$ *by 1 and relations (6.8) and (6.9) hold with a replacement of* $j_n!$ *by 1.*

Proof. Assume that $u = v = 1$ and $P(Y = 1) = p > 0$.

It is not difficult to check that

$$Eg(X)I_{\{Y=1\}} = pEg(\bar{X}) \tag{6.10}$$

for every bounded Borel function $g(x)$.

For sake of brevity, we prove relation (6.7) for $n = 3$. The proof for another n follows the same pattern.

Put $I_j = I_{\{Y_j=1\}}$ for $j = 1, 2, 3$. By the Fubini theorem, we get

$$\begin{aligned} P &= P\left(X_1 + X_2 + X_3 < x,\ I_{\{Y_1=Y_2=Y_3=1\}} = 1\right) \\ &= EI_{\{X_1+X_2+X_3<x\}}I_1I_2I_3 \\ &= EI_3\left[EI_2\left(EI_1I_{\{X_1+x_2+x_3<x\}}\right)_{x_2=X_2}\right]_{x_3=X_3}. \end{aligned}$$

Applying (6.10) three times, we obtain

$$\begin{aligned} P &= pEI_3\left[EI_2\left(EI_{\{\bar{X}_1+x_2+x_3<x\}}\right)_{x_2=X_2}\right]_{x_3=X_3} \\ &= p^2EI_3\left[E\left(EI_{\{\bar{X}_1+x_2+x_3<x\}}\right)_{x_2=\bar{X}_2}\right]_{x_3=X_3} \\ &= p^3E\left[E\left(EI_{\{\bar{X}_1+x_2+x_3<x\}}\right)_{x_2=\bar{X}_2}\right]_{x_3=\bar{X}_3}. \end{aligned}$$

Using again the Fubini theorem, we have

$$P\left(X_1 + X_2 + X_3 < x,\ I_1I_2I_3 = 1\right) = p^3P(\bar{X}_1 + \bar{X}_2 + \bar{X}_3 < x).$$

By $x > 0$, we get

$$P\left((X_1 + X_2 + X_3)I_1I_2I_3 < x\right) = P\left(X_1 + X_2 + X_3 < x,\ I_1I_2I_3 = 1\right).$$

This implies relation (6.7) with 1 instead of $n!$.

Further, we have

$$\begin{aligned} P(M_n \geq x) &\leq \sum_{k=1}^{n-j_n} P\left((S_{k+j_n} - S_k)I_{\{Y_{k+1}=\cdots=Y_{k+j_n}=1\}} \geq x\right) \\ &\leq np^{j_n}P\left(\bar{S}_{j_n} \geq x\right), \end{aligned}$$

and we obtain(6.8) with 1 instead of $j_n!$.

Taking into account that $1 - x \leq e^{-x}$ for $x \geq 0$, we have

$$\begin{aligned} P(M_n < x) &\leq P\left(\bigcap_{m=0}^{[n/j_n]-1}\left\{(S_{(m+1)j_n} - S_{mj_n})I_{\{Y_{mj_n+1}=\cdots=Y_{(m+1)j_n}=1\}} < x\right\}\right) \\ &= \left(P\left(S_{j_n}I_{\{Y_1=\cdots=Y_{j_n}=1\}} < x\right)\right)^{[n/j_n]-1} \\ &= \left(1 - P\left(S_{j_n}I_{\{Y_1=\cdots=Y_{j_n}=1\}} \geq x\right)\right)^{[n/j_n]-1} \\ &\leq \exp\left\{-\left(\left[\frac{n}{j_n}\right] - 1\right)P(S_{j_n}I_{\{Y_1=\cdots=Y_{j_n}=1\}} \geq x)\right\} \\ &\leq \exp\left\{-\left[\frac{n}{j_n}\right]p^{j_n}P(\bar{S}_j \geq x)\right\} \end{aligned}$$

and we get (6.9) with 1 instead of $j_n!$.

For the case $u < v$ and $P(Y = y) = 0$ for all y, we have

$$P\left(S_n I_{\{u \le Y_1 \le \cdots \le Y_n \le v\}} \ge x\right) =$$
$$\frac{1}{n!} P\left(S_n I_{\{Y_1 \in [u,v], \ldots Y_n \in [u,v]\}} \ge x\right) = \frac{p^n}{n!} P(\bar{S}_n \ge x).$$

In the last equality, we have used relation (6.7) for sums over head runs which has been proved above. Now relations (6.8) and (6.9) follow from (6.7) by the same way as before. □

Turn to the proof of Theorem 6.1.

Proof. Start with the case A) when $u = v = 1$ and $P(Y = 1) = p \in (0, 1)$.

If $\omega < \infty$ and either $(\tau_n - \log l_n)/i_n \ge 1/c_0$ (for $c_0 > 0$), or $(\tau_n - \log l_n)/i_n \to \infty$ (for $c_0 = 0$), then $b_n \sim i_n \omega$ and inequality (6.6) holds obviously. Hence, we concern with the case when the argument of $\gamma(x)$ is from the corresponding interval.

Suppose first that $\rho_n = \tau_n / \log l_n \to \infty$. Put

$$N_{ij} = \left\{n : 2^j \le n < 2^{j+1},\ i_n = i\right\}, \quad n_{ij} = \min\left\{n : n \in N_{ij}\right\}. \tag{6.11}$$

Take $\varepsilon > 0$. Using the concavity of $\gamma(x)$, Lemmas 6.3 and 6.5 and relation (6.5), we have

$$\begin{aligned} P_{ij} &= P\left(\max_{n \in N_{ij}} \frac{M_n}{b_n} \ge 1 + \varepsilon\right) \\ &\le 2^{j+1} p^i\, P\left(\bar{S}_i \ge (1+\varepsilon)\, i\, \gamma\left(\frac{\tau_{n_{ij}} + \log l_{n_{ij}}}{i}\right)\right) \\ &\le 2^{j+1} p^i \exp\left\{-(1+\varepsilon)(\tau_{n_{ij}} + \log l_{n_{ij}})\right\} \\ &\le 2 \exp\left\{-\varepsilon \tau_{n_{ij}} - (1+\varepsilon) \log l_{n_{ij}})\right\} \le \exp\left\{-3 \log l_{2^j}\right\} \end{aligned}$$

for all sufficiently large j. We have used $\rho_n \to \infty$ in the last inequality. Since $m_j = \#\{i : N_{ij} \ne \emptyset\} \le i_{2^j} \le l_{2^j}$, we have

$$P_j = P\left(\max_{2^j \le n < 2^{j+1}} \frac{M_n}{b_n} \ge 1 + \varepsilon\right) \le \frac{C}{j^2}$$

for all sufficiently large j and some positive constant C. Hence, the series $\sum_j P_j$ converges. By the Borel–Cantelli lemma, we get (6.6).

Assume now that $\rho_n \le a < \infty$. Take $\theta > 1$. For $i \ge 2$ and $j \ge 0$, put

$$N_{ij} = \left\{n : \theta^j \le 1 + \rho_n < \theta^{j+1},\ i_n = i\right\}, \quad n_{ij} = \max\left\{n : n \in N_{ij}\right\}. \tag{6.12}$$

Take $\varepsilon > 0$. Applying again Lemmas 6.3 and 6.5 and relation (6.5), we obtain

$$R(n) = P\left(M_n \geq (1+\varepsilon)b_n\right) \leq np^{i_n} P\left(\bar{S}_{i_n} \geq (1+\varepsilon)b_n\right)$$
$$\leq C \exp\left\{-(1+\varepsilon+\varepsilon\rho_n)\log i_n\right\}$$

for all sufficiently large n, where C is a positive constant. It follows that the series $\sum_i \sum_j R(n_{ij})$ converges. By the Borel–Cantelli lemma, we have

$$\limsup_{i\to\infty} \frac{M_{n_{ij}}}{b_{n_{ij}}} \leq 1+\varepsilon \quad \text{a.s.}$$

uniformly over j. By the concavity of the function $\gamma(x)$, the inequality $b_n \geq \theta^{-3} b_{n_{ij}}$ holds for all $n \in N_{ij}$, all sufficiently large i and all j. Moreover, $M_n \leq M_{n_{ij}}$ for $n \in N_{ij}$. It yields that lim sup in (6.6) is less or equal to $(1+\varepsilon)\theta^3$. This implies (6.6).

Note that if $i_n = c$, then the bound for $R(n)$ holds with l_n instead of i_n and the rest of proof is the same as before.

Assume finally that there exist sequences $\{n'_k\}$ and $\{n''_k\}$ of natural numbers such that $\{n'_k\} \cup \{n''_k\} = \mathbf{N}$, $\rho_{n'_k} \to \infty$ as $k \to \infty$ and $\rho_{n''_k} \leq a$. In the same way as before, we separately prove (6.6) for each of these sequences. From this, we obtain (6.6) for all natural numbers.

The proof of Theorem 6.1 is completed in case A). We deal with the case B) in the similar way. Of course, we exclude again the cases in which $b_n \sim i_n\omega$ and (6.6) is obvious.

Suppose first that $\rho_n = \tau_n / \log l_n \to \infty$ and define N_{ij} and n_{ij} by (6.11) again.

Take $\varepsilon > 0$. By (6.8), the properties of the function $\gamma(x)$ and Lemmas 6.3 and 6.5, we have

$$P_{ij} = P\left(\max_{n\in N_{ij}} \frac{M_n}{b_n} \geq 1+\varepsilon\right) \leq \frac{2^{j+1}}{i!} p^i \exp\left\{-(1+\varepsilon)(\tau_{n_{ij}} + \log l_{n_{ij}})\right\}$$
$$\leq C \exp\left\{-\varepsilon\tau_{n_{ij}} - (1+\varepsilon)\log l_{n_{ij}}\right\}$$

for all sufficiently large j and some positive constant C. The remainder of the proof is the same as for head runs.

Assume now that $\rho_n \leq a < \infty$. Take $\theta > 1$. For $i \geq 2$ and $j \geq 0$, define N_{ij} and n_{ij} by (6.12) again.

Take $\varepsilon > 0$. Applying Lemmas 6.3 and 6.5 and relation (6.5), we have

$$R(n) = P\left(M_n \geq (1+\varepsilon)b_n\right) \leq \frac{np^{i_n}}{i_n!} P\left(\bar{S}_{i_n} \geq (1+\varepsilon)b_n\right)$$
$$\leq C \exp\left\{-(1+\varepsilon+\varepsilon\rho_n)\log i_n\right\}$$

for all sufficiently large n, where C is a positive constant.

The remainder of the proof repeats that for head runs. We omit details. □

Theorem 6.1 gives an upper bound for the rate of the growth of M_n. Turn to the lower bounds. The first result is as follows.

Theorem 6.2. *Assume that $\bar{X}$ is non-degenerate, $E\bar{X} \geq 0$ and $h_0 > 0$. Suppose that $i_n \geq 1$, i_n is non-decreasing and one of two following conditions holds:*

1) $h_n = f^{-1}(\tau_n/i_n) < h_0$ and

$$h_n \sigma(h_n) = o(\sqrt{i_n} f(h_n)). \tag{6.13}$$

2) $\tau_n/i_n \to \infty$.

Then

$$\limsup \frac{M_n}{c_n} \geq 1 \quad a.s., \tag{6.14}$$

where $c_n = i_n \gamma(\tau_n/i_n)$.

The next remark follows from Lemma 2.25.

Remark 6.1. If $h_n \leq \tilde{h} < h_0$ for some $\tilde{h} < \infty$ and $\tau_n \to \infty$, then (6.13) holds.

Proof. Assume first that condition A) holds, i.e. $u = v = 1$ and $P(Y = 1) = p \in (0,1)$.

Suppose that condition 1) holds. Put

$$R_n = (S_n - S_{n-i_n}) I_{\{Y_{n-i_n+1} = \cdots = Y_n = 1\}}.$$

Take $\varepsilon \in (0, 1/2)$. By Lemmas 6.4 and 6.5 and relation (6.5), we have

$$\begin{aligned} P(n) &= P\left(R_n \geq (1-2\varepsilon)c_n\right) = p^{i_n} P\left(\bar{S}_{i_n} \geq (1-2\varepsilon)c_n\right) \\ &\geq p^{i_n} \exp\{-(1-\varepsilon)\tau_n\} \geq \frac{1}{n} \end{aligned}$$

for all sufficiently large n.

Put $n_k = ak \log k$ for $k \geq 3$, where a is chosen such that $n_{k+1} - i_{n_{k+1}} > n_k$. Then the series $\sum_k P(n_k)$ diverges and the events $\{R_{n_k} \geq (1-2\varepsilon)c_{n_k}\}$ are independent. Applying the Borel–Cantelli lemma and the inequality $M_n \geq R_n$, we obtain (6.14).

Assume that condition 2) holds. Put $c(n) = \tau_n/i_n$. For all $\varepsilon > 0$, we have

$$Q(n) = P\left(M_n < (1+\varepsilon)^{-3}b_n\right)$$
$$\le \exp\left\{-\frac{np^{i_n}}{2i_n}\left(P\left(\bar{X} \ge (1+\varepsilon)^{-3}\gamma(c(n))\right)\right)^{i_n}\right\}.$$

By the method used in Lemma 2.5 from [Mason (1989)], we construct the sequence of natural numbers $\{n_r\}$ such that

$$\left(P\left(\bar{X} \ge (1+\varepsilon)^{-3}\gamma(c(n_r))\right)\right)^{\log n_r/c(n_r)} \ge \exp\left\{-(1+\varepsilon)^{-1}\log n_r\right\} \quad (6.15)$$

and $n_r \ge r$ for all sufficiently large r.

Then we get

$$Q(n_r) \le \exp\{-Cn_r^{\delta}\} \quad (6.16)$$

for some $\delta > 0$ and all sufficiently large r. Hence, the series $\sum_r Q(n_r)$ converges. Applying of the Borel–Cantelli lemma yields (6.14).

The proof is completed in case A). Turn to the case B).

Assume that condition 1) holds. Put

$$R_n = (S_n - S_{n-i_n})I_{\{u \le Y_{n-i_n+1} \le \cdots \le Y_n \le v\}}.$$

Take $\varepsilon \in (0, 1/2)$. By Lemmas 6.4 and 6.5 and relation (6.5), we have

$$P(n) = P(R_n \ge (1-2\varepsilon)c_n) = \frac{p^{i_n}}{i_n!}P\left(\bar{S}_{i_n} \ge (1-2\varepsilon)c_n\right)$$
$$\ge \frac{p^{i_n}}{i_n!}\exp\{-(1-\varepsilon)\tau_n\} \ge \frac{1}{n}$$

for all sufficiently large n. In the same way as for the case A), we again arrive at (6.14).

Assume that condition 2) holds. Put $c(n) = \tau_n/i_n$. Remember that $c(n)$ is quite different in this case. Take $\varepsilon > 0$. By (6.9), we have

$$Q(n) = P\left(M_n < (1+\varepsilon)^{-3}b_n\right)$$
$$\le \exp\left\{-\left[\frac{n}{i_n}\right]\frac{p^{i_n}}{i_n!}P\left(\bar{S}_{i_n} \ge (1+\varepsilon)^{-3}i_n\gamma(c(n))\right)\right\}$$
$$\le \exp\left\{-\left[\frac{n}{i_n}\right]\frac{p^{i_n}}{i_n!}P\left(\bar{X} \ge (1+\varepsilon)^{-3}\gamma(c(n))\right)^{i_n}\right\}.$$

By the same method as used in Lemma 2.5 from [Mason (1989)], we construct the sequence of natural numbers $\{n_r\}$ such that (6.15) holds and $n_r \ge r$ for all sufficiently large r. An application of the Stirling formula, the definition of l_n and (6.15) yields the similar bound for $Q(n_r)$ as before. It follows that the series $\sum_r Q(n_r)$ converges and by the Borel–Cantelli lemma, we obtain (6.14) again. □

We now investigate a minimal rate of the growth of M_n.

Theorem 6.3. *Assume that $\bar{X}$ is non-degenerate, $E\bar{X} \geq 0$, $h_0 > 0$, $i_n \geq 1$, i_n is non-decreasing and*

$$\liminf \frac{\tau_n}{\log l_n} > 1. \tag{6.17}$$

Assume that one of three following condition hold:

1) $h_n = f^{-1}(\tau_n/i_n) < h_0$ and (6.13) holds.

2) $\tau_n/i_n \to \infty$ and $\omega < \infty$.

3) $\tau_n/i_n \to \infty$ and

$$\lim_{z\to\infty} \frac{\gamma(-\log(1-F(z)))}{z} = 1\,. \tag{6.18}$$

Then

$$\liminf \frac{M_n}{d_n} \geq 1 \quad a.s., \tag{6.19}$$

where $d_n = i_n\gamma((\tau_n - \log l_n)/i_n)$.

According to [Mason (1989)], condition (6.18) holds for the normal, geometric, Poisson and Weibull distributions.

Proof. Start with the case A).

Assume that condition 1) holds.

Suppose first that $\rho_n = \tau_n/\log l_n \to \infty$. Take $\varepsilon \in (0, 1/2)$. By Lemmas 6.4 and 6.5 and relation (6.5), we have

$$\begin{aligned} Q(n) = P(M_n < (1-2\varepsilon)d_n) &\leq \exp\left\{-\frac{np^{i_n}}{2i_n}P\left(\bar{S}_{i_n} \geq (1-2\varepsilon)d_n\right)\right\} \\ &\leq \exp\left\{-\frac{3np^{i_n}}{8i_n}\exp\{-(1-\varepsilon)(\tau_n - \log l_n)\}\right\} \leq \exp\left\{-i_n^{\varepsilon(\rho_n-1)}\right\} \end{aligned} \tag{6.20}$$

for all sufficiently large n. It implies that the series $\sum_n Q(n)$ converges and an application of the Borel–Cantelli lemma yields (6.19).

Suppose now that $\rho_n \leq a$. Take $\theta > 1$. For $i \geq 2$ and $j \geq -1$, put

$$\begin{aligned} N_{i,-1} &= \{n : \delta < \rho_n - 1 < 1,\ i_n = i\}, \\ N_{ij} &= \left\{n : \theta^j \leq \rho_n - 1 < \theta^{j+1},\ i_n = i\right\}, \quad m_{ij} = \min\{n : n \in N_{ij}\}\,. \end{aligned}$$

Hence, the series $\sum_j \sum_k Q(m_{jk})$ converges and by the Borel–Cantelli lemma, we have

$$\liminf_{i\to\infty} \frac{M_{n_{ij}}}{d_{n_{ij}}} \geq 1 - 2\varepsilon \quad \text{a.s.}$$

uniformly over j. Making use of the concavity of the function $\gamma(x)$, we conclude that $d_n \leq \theta^3 d_{n_{ij}}$ for $n \in N_{ij}$, all sufficiently large i and all j. From the last inequality and the inequality $M_n \geq M_{n_{ij}}$ for $n \in N_{ij}$, it follows that lim inf in (6.19) is greater than $(1-2\varepsilon)\theta^{-3}$. This yields (6.19).

Relation (6.19) for oscillating ρ_n can be proved in the same way as before in Theorem 6.1.

Assume that condition 2) holds. Take $0 < \delta < 1$. We have again that

$$Q(n) = P(M_n < \delta\omega i_n) \leq \exp\left\{-\frac{np^{i_n}}{2i_n}\left(P(\bar{X} \geq \delta_1\omega)\right)^{i_n}\right\}.$$

In the same way as before, we prove that the series $\sum_n Q(n)$ converges. An application of the Borel–Cantelli lemma completes the proof of (6.19) in this case.

Assume that condition 3) is satisfied. Put $c(n) = \tau_n/i_n$. For all $\varepsilon > 0$, we have

$$\begin{aligned} Q(n) &= P(M_n < (1+\varepsilon)^{-2} b_n) \\ &\leq \exp\left\{-\frac{np^{i_n}}{2i_n}\left(P(\bar{X} \geq (1+\varepsilon)^{-2}\gamma(c(n)))\right)^{i_n}\right\}. \end{aligned}$$

Making use of the inequality

$$P\left(\bar{X} \geq (1+\varepsilon)^{-2}\gamma(c(n))\right) \geq \exp\left\{-(1+\varepsilon)^{-1}c(n)\right\}$$

(see [Mason (1989)], p. 264) we prove that the series $\sum_n Q(n)$ converges. Together with the Borel–Cantelli lemma, this completes the proof of Theorem 6.2 in case A).

Turn to the case B).

Assume that condition 1) holds. Define $Q(n)$ as in the proof for the case A) under condition 1). In the same way as before, we prove bound (6.20). The remainder part of the proof is the same as before.

Suppose that condition 2) holds. Since $\omega < \infty$, we can assume that $b_n = i_n\omega$. Define $Q(n)$ as in the proof for the case A) under condition 2). In the same way as before, one can check that the series $\sum_n Q(n)$ converges. Applying of the Borel–Cantelli lemma completes the proof in this case.

Assume that condition 3) holds. Define $Q(n)$ as in the proof for the case A) under condition 3). In the same way as before, one can check that the series $\sum_n Q(n)$ converges. Making use of the Borel–Cantelli lemma, we arrive at the desired conclusion in this case.

The proof of Theorem 6.3 is completed. □

Theorems 6.1–6.3 imply the following result which gives universal strong laws for increments of sums of i.i.d. random variables over head runs and monotone blocks.

Theorem 6.4. *If the conditions of Theorem 6.2 hold and*

$$\lim \frac{\tau_n}{\log l_n} = \infty, \tag{6.21}$$

then

$$\limsup \frac{M_n}{b_n} = 1 \quad a.s. \tag{6.22}$$

If the conditions of Theorem 6.3 and relation (6.21) hold, then

$$\lim \frac{M_n}{b_n} = 1 \quad a.s. \tag{6.23}$$

Note that $\liminf \tau_n / i_n > 0$ implies (6.21) and (6.17). It is not difficult to check that $b_n \sim c_n \sim d_n$ provided (6.21) holds.

6.4 Corollaries of the Universal Strong Laws

In this section, we discuss various corollaries from the results of the previous section.

First of all, note that the norming sequences in the universal laws can easily be calculated for the completely asymmetric stable laws and the normal law. Remember that for the stable d.f. F with c.f. (2.2), we have

$$\gamma(x) = \left(\frac{\alpha x}{\alpha - 1}\right)^{(\alpha-1)/\alpha}$$

for all $x > 0$. If $\alpha = 2$, then F is the standard normal d.f. and $\gamma(x) = \sqrt{2x}$.

It implies that the norming c_n for the stable laws is as follows:

$$c_n = i_n^{1/\alpha} \left(\frac{\alpha \tau_n}{\alpha - 1}\right)^{(\alpha-1)/\alpha}.$$

It yields that $c_n = (2 i_n \tau_n)^{1/2}$, in the Gaussian case. Formulae for b_n and d_n follow from the last formulae by a replacement of τ_n by $\tau_n \pm \log l_n$ correspondingly.

We now state strong limit theorems for increments of sums over head runs and monotone blocks. It turns out in these cases that the type of asymptotic behaviour of M_n depends on asymptotics of ratios τ_n and i_n. If this ratio is separated from zero, then we get the Erdős–Rényi laws. If this ratio tends to zero, then we arrive at the Csörgő–Révész laws.

We start with the Erdős–Rényi law for a general situation. Further, we deal with the cases of head runs and monotone blocks separately.

Theorem 6.5. *Assume that $\bar{X}$ is non-degenerate, $E\bar{X} \geq 0$ and $h_0 > 0$. Assume that $i_n \geq 1$, i_n is non-decreasing and*

$$\liminf \frac{\tau_n}{i_n} > 0. \tag{6.24}$$

Then (6.22) holds with $b_n = i_n \gamma((1-\delta_n)\log n/i_n)$.
If additionally $\omega < \infty$ or (6.18) holds, then (6.23) holds.

The Erdős–Rényi law for increments of sums over head runs is as follows.

Theorem 6.6. *Assume that $\bar{X}$ is non-degenerate, $E\bar{X} \geq 0$ and $h_0 > 0$, Assume that $i_n \geq 1$, i_n is non-decreasing and condition A) holds.*

If $\delta_n \to B \in [0,1)$, then (6.24) is satisfied and, consequently, the result of Theorem 6.5 holds. If $i_n \to \infty$, then $b_n \sim i_n \gamma((1-\delta_n)\log(1/p)/\delta_n)$ and $b_n \sim B l_n \gamma((1-B)\log(1/p)/B)$ for $B > 0$.

The next theorem is the corresponding result for increments of sums over monotone blocks.

Theorem 6.7. *Assume that $\bar{X}$ is non-degenerate, $E\bar{X} \geq 0$ and $h_0 > 0$, Assume that $i_n \geq 1$, i_n is non-decreasing and condition B) holds.*

If $(1-\delta_n)\log\log n \to \infty$, then (6.24) is satisfied and, consequently, the result of Theorem 6.5 holds. If $i_n \to \infty$, then $b_n \sim i_n \gamma((1-\delta_n)\log\log n/\delta_n)$. If $(1-\delta_n)\log\log n \to B + \log(1/p)$ with $B \in (0, 1/c_0)$, then (6.24) holds and $b_n \sim l_n \gamma(B)$.

We see that the results of Theorems 6.6 and 6.7 are quite different. We have the Erdős–Rényi law in the case of head runs when δ_n is separated from one. In the case of monotone blocks, δ_n can tend to one with an appropriate rate of convergence.

Turn to the Csörgő–Révész laws. This case arises when the arguments of the function $\gamma(x)$ in the definitions of b_n, c_n and d_n tend to zero. Under various additional conditions on the distribution of $\bar{X}$, the asymptotics of $\gamma(x)$ at zero is known from above considerations. This yields simpler formulae for norming sequences. We separately consider the case of finite variations, the case of non-normal attraction to the normal law and the cases of normal and non-normal attraction to completely asymmetric stable laws.

Assume first that $E\bar{X}^2 < \infty$. Then we have the next general result.

Theorem 6.8. *Assume that* $E\bar{X} = 0$, $E\bar{X}^2 = 1$ *and* $h_0 > 0$. *Assume that* i_n *is non-decreasing and*

$$\lim \frac{\tau_n}{i_n} = 0. \tag{6.25}$$

Then (6.6) hods with $b_n = (2l_n(\tau_n + \log l_n))^{1/2}$. *If additionally* $\tau_n \to \infty$, *then (6.14) hold with* $c_n = (2l_n\tau_n)^{1/2}$. *If additionally (6.17) is satisfied, then (6.19) holds with* $d_n = (2l_n(\tau_n - \log l_n))^{1/2}$. *If additionally (6.21) is satisfied, then (6.23) holds with* $b_n = (2l_n\tau_n)^{1/2}$.

Note that condition (6.25) fails when $\{\delta_n\}$ is separated from 1. Hence, one can assume that $\delta_n \to 1$. It follows that $i_n \sim l_n$ and one can replace i_n by l_n in the normalizing sequences. Hence, one can assume without loss of generality that $i_n \geq 1$ for all n.

For increments over head runs, we obtain the following result.

Theorem 6.9. *Assume that* $E\bar{X} = 0$, $E\bar{X}^2 = 1$ *and* $h_0 > 0$. *Assume that condition A) holds and* i_n *is non-decreasing.*

If $\delta_n \to 1$, *then (6.25) is satisfied and, consequently, the result of Theorem 6.8 holds.*

For increments over monotone blocks, we arrive at the next result.

Theorem 6.10. *Assume that* $E\bar{X} = 0$, $E\bar{X}^2 = 1$ *and* $h_0 > 0$. *Assume that condition B) holds and* i_n *is non-decreasing.*

If $(1-\delta_n)\log l_n \searrow \log(1/p)$, *then (6.25) is fulfilled, the result of Theorem 6.8 holds and* $\tau_n \sim (l_n - i_n)\log l_n - i_n \log(1/p)$. *For* $p < 1$, *we have* $\tau_n \sim l_n(r_n + (\log(1/p))^2/\log l_n)$, *where* $r_n \searrow 0$, *and, consequently,* $b_n \sim c_n \sim d_n$. *The last relation can fail for* $p = 1$.

We see from Theorems 6.9 and 6.10 that the behaviour of increments of sums of i.i.d. random variables over monotone blocks is much more complicated.

Two last theorems yield variants of the LIL. In the case of head runs, if $\tau_n \sim c\log l_n$, then relations (6.6) and (6.14) imply the LIL with the norming sequence $(2l_n \log l_n)^{1/2}$ which is equivalent to $(2\log n \log\log n/\log(1/p))^{1/2}$. In the case of monotone blocks with $p = 1$ and $\tau_n \sim c\log l_n$, we also get the LIL with the norming sequence is $(2l_n \log l_n)^{1/2}$, but it is equivalent to $(2\log n)^{1/2}$. Unfortunately, the exact constant can not be specified since b_n

and c_n distinguish. Moreover, we have LIL for the case of monotone blocks provided $p = 1$.

We then arrive at the following result for the case of head runs.

Corollary 6.1. *Assume that conditions of Theorem 6.9 hold. If $\tau_n \sim c \log l_n$, then*

$$\limsup \frac{M_n}{\sqrt{2 \log n \log \log n / \log(1/p)}} = L \quad a.s.,$$

where $\sqrt{c} \leq L \leq \sqrt{c+1}$.

The LIL for the case of monotone blocks is as follows.

Corollary 6.2. *Assume that conditions of Theorem 6.10 hold. If $p = 1$ and $\tau_n \sim c \log l_n$, then*

$$\limsup \frac{M_n}{\sqrt{2 \log n}} = L \quad a.s.$$

where $\sqrt{c} \leq L \leq \sqrt{c+1}$.

Turn to the case $E\bar{X}^2 = \infty$.

We start with the domain of non-normal attractions of the normal law. We have the following result.

Theorem 6.11. *Assume that $E\bar{X} = 0$, $h_0 > 0$ and $F \in D(2)$. Put*

$$\hat{m}(h) = hG\left(\frac{1}{h}\right), \quad \hat{f}(h) = \frac{h^2}{2} G\left(\frac{1}{h}\right), \quad where \quad G(x) = \int_{-x}^{0} u^2 dF(u).$$

Suppose that i_n is non-decreasing and (6.25) holds.

Then (6.6) holds with $b_n = i_n \hat{m}(\hat{f}^{-1}((\tau_n + \log l_n)/i_n))$. If additionally $\tau_n \to \infty$, then (6.14) holds with $c_n = i_n \hat{m}(\hat{f}^{-1}(\tau_n/i_n))$. If additionally (6.17) is satisfied, then (6.19) holds with $d_n = i_n \hat{m}(\hat{f}^{-1}((\tau_n - \log l_n)/i_n))$. If additionally (6.21) is satisfied, then (6.23) holds with $b_n = i_n \hat{m}(\hat{f}^{-1}(\tau_n/i_n))$.

Our next result is for $\bar{X}$ from the domain of the normal attraction of the completely asymmetric stable laws.

Theorem 6.12. *Assume that $E\bar{X} = 0$, $h_0 > 0$ and $F \in DN(\alpha)$, $\alpha \in (1, 2)$. Suppose that i_n is non-decreasing and (6.25) holds. Put $\lambda = (\alpha - 1)/\alpha$.*

Then (6.6) holds with $b_n = \lambda^{-\lambda} l_n^{1/\alpha} (\tau_n + \log l_n)^{\lambda}$. If additionally $\tau_n \to \infty$, then (6.14) holds with $c_n = \lambda^{-\lambda} l_n^{1/\alpha} \tau_n^{\lambda}$. If additionally (6.17) is satisfied, then (6.19) holds with $d_n = \lambda^{-\lambda} l_n^{1/\alpha} (\tau_n - \log l_n)^{\lambda}$. If additionally (6.21) is satisfied, then (6.23) holds with $b_n = \lambda^{-\lambda} l_n^{1/\alpha} \tau_n^{\lambda}$.

For increments over head runs, we obtain the following result.

Theorem 6.13. *Assume that $E\bar{X} = 0$, $h_0 > 0$ and $F \in DN(\alpha)$, $\alpha \in (1, 2)$. Suppose that condition A) holds, i_n is non-decreasing and (6.25) holds. Put $\lambda = (\alpha - 1)/\alpha$.*

If $\delta_n \to 1$, then (6.25) is satisfied and, consequently, the result of Theorem 6.12 holds.

For increments over monotone blocks, we have the next result.

Theorem 6.14. *Assume that $E\bar{X} = 0$, $h_0 > 0$ and $F \in DN(\alpha)$, $\alpha \in (1, 2)$. Suppose that condition B) holds, i_n is non-decreasing and (6.25) holds. Put $\lambda = (\alpha - 1)/\alpha$.*

If $(1-\delta_n)\log l_n \searrow \log(1/p)$, then (6.25) is fulfilled, the result of Theorem 6.12 holds and the behaviour of τ_n is the same as in Theorem 6.10.

Turn to analogues of LIL for increments of sums over head runs and monotone blocks in the case of domains of normal attraction of completely asymmetric stable laws.

In the case of head runs, if $\tau_n \sim c \log l_n$, then the norming sequence is $\lambda^{-\lambda} l_n^{1/\alpha} (\log l_n)^{\lambda} \sim \lambda^{-\lambda} \log(1/p)^{-1/\alpha} (\log n)^{1/\alpha} (\log\log n)^{\lambda}$. In the case of monotone blocks with $p = 1$ and $\tau_n \sim c \log l_n$, the norming sequence is $\lambda^{-\lambda} l_n^{1/\alpha} (\log l_n)^{\lambda} \sim \lambda^{-\lambda} (\log n)^{1/\alpha} (\log\log n)^{\lambda - 1/\alpha}$.

We then get the next result in the case of head runs.

Corollary 6.3. *Assume that conditions of Theorem 6.13 hold. If $\tau_n \sim c \log l_n$, then*

$$\limsup \frac{M_n}{\lambda^{-\lambda} \log(1/p)^{-1/\alpha} (\log n)^{1/\alpha} (\log\log n)^{\lambda}} = L \quad a.s.,$$

where $c^{\lambda} \le L \le (1+c)^{\lambda}$.

The LIL for the case of monotone blocks is as follows.

Corollary 6.4. *Assume that conditions of Theorem 6.14 hold. If $p = 1$ and $\tau_n \sim c \log l_n$, then*

$$\limsup \frac{M_n}{\lambda^{-\lambda} (\log n)^{1/\alpha} (\log\log n)^{\lambda - 1/\alpha}} = L \quad a.s.,$$

where $c^{\lambda} \le L \le (1+c)^{\lambda}$.

Turn to the case of non-normal attraction to completely asymmetric stable laws.

Theorem 6.15. *Assume that* $E\bar{X} = 0$, $h_0 > 0$ *and* $F \in D(\alpha)$, $\alpha \in (1,2)$. *Put*

$$\hat{m}(h) = \alpha\frac{\Gamma(2-\alpha)}{\alpha-1}h^{\alpha-1}G\left(\frac{1}{h}\right), \quad \hat{f}(h) = \Gamma(2-\alpha)h^{\alpha}G\left(\frac{1}{h}\right),$$

where $G(x) = x^{\alpha}F(-x)$, $x > 0$. *Suppose that* i_n *is non-decreasing and* (6.25) *holds.*

Then (6.6) *holds with* $b_n = i_n\hat{m}(\hat{f}^{-1}((\tau_n + \log l_n)/i_n))$. *If additionally* $\tau_n \to \infty$, *then* (6.14) *holds with* $c_n = i_n\hat{m}(\hat{f}^{-1}(\tau_n/i_n))$. *If additionally* (6.17) *is satisfied, then* (6.19) *holds with* $d_n = i_n\hat{m}(\hat{f}^{-1}((\tau_n - \log l_n)/i_n))$. *If additionally* (6.21) *is satisfied, then* (6.23) *holds with* $b_n = i_n\hat{m}(\hat{f}^{-1}(\tau_n/i_n))$.

We finally concern with the SLLN which follows from Theorem 6.4.

Theorem 6.16. *Assume that* $\bar{X}$ *is non-degenerate,* $E\bar{X} \geq 0$ *and* $h_0 > 0$. *If* i_n *is non-decreasing and relations* (6.25) *and* (6.21) *hold, then*

$$\lim \frac{M_n}{l_n} = E\bar{X} \quad a.s.$$

Proof. If $E\bar{X} > 0$, then the result immediately follows from Theorem 6.4. and the equality $\gamma(0) = E\bar{X}$.

Assume that $E\bar{X} = 0$. Take $\varepsilon > 0$. We have

$$M_n' - \varepsilon i_n \leq M_n \leq M_n'$$

where

$$M_n' = \max_{0 \leq k \leq n - i_n}(S_{k+i_n} - S_k + \varepsilon i_n)I_{\{u \leq Y_{k+1} \leq \cdots \leq Y_{k+i_n} \leq v\}}.$$

By the first part of the proof, we get

$$\lim \frac{M_n'}{l_n} = \varepsilon \quad \text{a.s.}$$

Since ε is an arbitrary positive number, the result follows from the last relation and the above inequalities. $\square$

6.5 Bibliographical Notes

The a.s. asymptotic behaviour of the longest head runs has been considered in [Feller (1971)], [Erdős and Révész (1975)] and [Deheuvels (1985)]. Various generalizations of these results may be found in [Frolov and Martikainen (1999)] and references therein.

The a.s. limiting behaviour of monotone blocks has been studied in [Pittel (1981)], [Révész (1983)], [Novak (1992)]. Generalizations of these results may be found in [Frolov and Martikainen (1999)] and references therein.

Results on the asymptotic behaviour of increasing runs in $\mathbb{R}^d$ and random fields may be found in [Frolov and Martikainen (1999)] and [Frolov and Martikainen (2001)]) correspondingly.

Results for increments of sums over head runs and monotone blocks have been proved in [Frolov *et al.* (1998)] , [Frolov (1999)], [Frolov *et al.* (2000a)] , [Frolov *et al.* (2000b)] and [Frolov (2001)].

Main results of this chapter are obtained in [Frolov (1999, 2001, 2003e)].

Bibliography

Amosova, N. N. (1972). On limit theorems for probabilities of moderate deviations, *Vestnik Leningrad Univ.* , 13, pp. 5–14, english transl.: Vestnik Leningrad Univ. Math., **5**, 197–210, (1978).

Amosova, N. N. (1979). On probabilities of moderate deviations for sums of independent random variables, *Theor. Probab. Appl.* **24**, 4, pp. 858–865, english transl.: Theor. Probab. Appl., **24**, no. 4, 856–863, (1980).

Amosova, N. N. (1980). On narrow zones of integral normal convergence, *Zap. Nauchn. Semin. LOMI* **97**, pp. 6–14, english transl.: J. Sov. Math., **24**, no. 5, 483–489, (1984).

Amosova, N. N. (1984). Probabilities of large deviations in the case of stable limit distributions, *Matem. Zametki* **35**, pp. 125–131, english transl.: Mathematical Notes, **35**, no. 1, 68–71, (1984).

Bacro, J.-N. and Brito, M. (1991). On Mason's extension of the Erdős–Rényi law of large numbers, *Statist. Probab. Lett.* **11**, pp. 43–47.

Bacro, J.-N., Deheuvels, P. and Steinebach, J. (1987). Exact convergence rates in Erdős–Rényi type theorems for renewal processes, *Ann. Inst. Henri Poincaré* **23**, pp. 195–207.

Bahadur, R. R. and Ranga Rao, R. (1960). On deviations of the sample mean, *Ann. Math. Statist.* **31**, 4, pp. 1015–1027.

Binswanger, K. and Embrechts, P. (1994). Longest runs in coin tossing, *Insur. Math. Econom.* **15**, pp. 139–149.

Book, S. A. (1975a). A version of the Erdős–Rényi law of large numbers for independent random variables, *Bull. Inst. Math. Acad. Sinica* **3**, 2, pp. 199–211.

Book, S. A. (1975b). An extension of the Erdős–Rényi new law of large numbers, *Proc. Amer. Math. Soc.* **48**, 2, pp. 438–446.

Book, S. A. and Shore, T. R. (1978). On large intervals in the Csörgő–Révész theorem on increments of a Wiener process, *Z. Wahrsch. Verw. Geb.* **46**, pp. 1–11.

Borovkov, A. A. (1964). Analisys of large deviations in boundary problems with arbitrary bounds. I, *Siberian Math. J.* **5**, 2, pp. 253–289.

Borovkov, A. A. and Borovkov, K. A. (2008). *Asymptotic analysis of random walks. Heavy-tailed distributions* (Cambridge University Press, New York).

Brieman, L. (1968). A delicate LIL for non-decreasing stable processes, *Ann. Math. Statist.* **39**, pp. 1818–1824.

Cai, Z. (1992). Strong approximation and improved Erdős–Rényi laws for sums of independent non-identically distributed random variables, *J. Hangzhou Univ.* **19**, 3, pp. 240–246.

Chan, A. H. (1976). Erdős–Rényi type modulus of continuity theorems for Brownian sheets, *Studia Sci. Math. Hungar.* **11**, pp. 59–68.

Chernoff, H. (1952). A measure of asymptotic efficiency for tests of a hypothesis based on the sum of observations, *Ann. Math. Statist.* **23**, 4, pp. 493–507.

Choi, Y. K. and Kôno, N. (1999). How big are increments of a two-parameter Gaussian processes, *J. Theoret. Probab.* **12**, pp. 105–129.

Cramér, H. (1938). Sur un nouveau théorème limite de la théorie des probabilités, *Actual. Sci. Indust.* , 736, pp. 5–23.

Csáki, E. and Révész, P. (1979). How big must be the increments of a Wiener process? *Acta Math. Acad. Sci. Hungar.* **33**, pp. 37–49.

Csörgő, M. and Révész, P. (1978). How big are the increments of a multiparameter Wiener process? *Z. Wahrsch. verw. Geb.* **42**, pp. 1–12.

Csörgő, M. and Révész, P. (1979). How big are the increments of a Wiener process? *Ann. Probab.* **7**, pp. 731–737.

Csörgő, M. and Révész, P. (1981). *Strong approximations in probability and statistics* (Akadémiai Kiadó, Budapest).

Csörgő, M. and Steinebach, J. (1981). Improved Erdős–Rényi and strong approximation laws for increments of partial sums, *Ann. Probab.* **9**, pp. 988–996.

Csörgő, S. (1979). Erdős–Rényi laws, *Ann. Statist.* **7**, pp. 772–787.

Daniels, H. E. (1954). Saddlepoint approximations in statistics, *Ann. Math. Statist.* **25**, 4, pp. 631–650.

Deheuvels, P. (1985). On the Erdős–Rényi theorem for random fields and sequences and its relationships with the theory of runs and spacings, *Z. Wahrsch. verw. Geb.* **70**, pp. 91–115.

Deheuvels, P. and Devroye, L. (1987). Limit laws of Erdős–Rényi–Shepp type, *Ann. Probab.* **15**, pp. 1363–1386.

Deheuvels, P. and Steinebach, J. (1989). Sharp rates for increments of renewal processes, *Ann. Probab.* **17**, pp. 700–722.

Einmahl, U. and Mason, D. M. (1996). Some universal results on the behaviour of increments of partial sums, *Ann. Probab.* **24**, pp. 1388–1407.

Embrechts, P. and Clüppelberg, C. (1993). Some aspects of insurance mathematics, *Theor. Probab. Appl.* **38**, 2, pp. 374–416, english transl.: Theor. Probab. Appl., **38**, no. 3, 262–295, (1993).

Erdős, P. and Rényi, A. (1970). On a new law of large numbers, *J. Analyse Math.* **23**, pp. 103–111.

Erdős, P. and Révész, P. (1975). On the length of the longest head-run, *In: Topics in Information Theory, Csiszár, I., Elias, P. (eds.). Amsterdam: North Holland,- Coll. Math. Soc. J. Bolyai* **16**, pp. 218–228.

Feller, W. (1943). Generalization of a probability limit theorem of Cramér, *Trans. Amer. Math. Soc.* **54**,, 3, pp. 361–372.

Feller, W. (1969). Limit theorems for probabilities of large deviations, *Z. Wahrsch. verw. Geb.* **14**, pp. 1–20.

Feller, W. (1971). *An introduction to probability theory and its applications, Vol. 1,2 (2nd edn.)* (Wiley, New York).

Fristedt, B. (1964). The behavior of increasing stable processes for both small and large times, *J. Math. Mech.* **13**, pp. 849–856.

Frolov, A. N. (1990). On the Erdős–Rényi law of large numbers when Cramér's condition is not fulfilled, *Vestnik Leningrad Univ., Ser. 1, i. 4* , 22, pp. 26–31, english transl.: Vestnik Leningrad Univ. Math., **23**, no. 4, 30–35, (1990).

Frolov, A. N. (1991). On the Erdős–Rényi law of large numbers for non-identically distributed random variables under violation of Cramér's condition, *Vestnik Leningrad Univ., Ser. 1, i. 2* , 8, pp. 57–61, english transl.: Vestnik Leningrad Univ. Math., **24**, no. 2, 66–70, (1991).

Frolov, A. N. (1993a). Exact convergence rate in the Erdős–Rényi law for non-identically distributed random variables, *Vestnik St.Petersburg Univ., Ser. 1, i. 1* , 1, pp. 124–125, english transl.: Vestnik St.Petersburg Univ. Math., **26**, no. 1, 60–61, (1993).

Frolov, A. N. (1993b). On asymptotic behaviour of increments of sums of independent random variables, *Vestnik St.Petersburg Univ., Ser. 1, i. 4* , 22, pp. 45–48, english transl.: Vestnik St.Petersburg Univ. Math., **26**, no. 4, 52–56, (1993).

Frolov, A. N. (1998). On one-sided strong laws for large increments of sums, *Statist. Probab. Lett.* **37**, pp. 155–165.

Frolov, A. N. (1999). On asymptotic behaviour of increments of sums over head runs, *Zap. Nauchn. Semin. POMI* **260**, pp. 263–277, english transl.: J. Math. Sci., **109**, no. 6, 2229–2240, (2002).

Frolov, A. N. (2000). On the asymptotic behaviour of increments of sums of independent random variables, *Doklady Akademii nauk* **372**, 5, pp. 596–599, english transl.: Doklady Math., **61**, no. 3, 409–412, (2000).

Frolov, A. N. (2001). On asymptotic behaviour of increments of sums over increasing runs, *Zap. Nauchn. Semin. POMI* **278**, pp. 248–260, english transl.: J. Math. Sci., **118**, no. 6, 5650–5657, (2003).

Frolov, A. N. (2002a). On asymptotic behaviour of large increments of sums of independent random variables, *Theor. Probab. Appl.* **47**, 2, pp. 366–374, english transl.: Theor. Probab. Appl., **47**, no. 2, 315–323, (2003).

Frolov, A. N. (2002b). On probabilities of moderate deviations of sums of independent random variables, *Zap. Nauchn. Semin. POMI* **294**, pp. 200–215, english transl.: J. Math. Sci., **127**, no. 1, 1787–1796, (2005).

Frolov, A. N. (2002c). One-sided strong laws for increments of sums of i.i.d. random variables, *Studia Sci. Math. Hungar.* **39**, pp. 333–359.

Frolov, A. N. (2002d). Strong limit theorems for increments of random fields, *Theory Stoch. Processes* **8 (24)**, Ns. 1–2, pp. 89–97.

Frolov, A. N. (2003a). Limit theorems for increments of processes with independent increments, *Doklady Akademii nauk* **393**, 2, pp. 165–169, english transl.: Doklady Math., **68**, no. 3, 345–349, (2003).

Frolov, A. N. (2003b). Limit theorems for increments of sums independent random variables, *Theor. Probab. Appl.* **48**, 1, pp. 104–121, english transl.: Theor. Probab. Appl., **48**, no. 1, 93–107, (2004).

Frolov, A. N. (2003c). On asymptotic behaviour of increments of random fields, *Zap. Nauchn. Semin. POMI* **298**, pp. 191–207, english transl.: J. Math. Sci., **128**, no. 1, 2604–2613, (2005).

Frolov, A. N. (2003d). On asymptotics of large increments of sums in non-i.i.d. case, *Acta Applicandae Mathematicae* **78**, pp. 129–136.

Frolov, A. N. (2003e). On asymptotics of the maximal gain without losses, *Statist. Probab. Lett.* **63**, pp. 13–23.

Frolov, A. N. (2003f). On the Erdős–Rényi law for renewal processes, *Teor. veroyatn. matem. statist.* **68**, pp. 142–151, (In Russian).

Frolov, A. N. (2003g). Strong limit theorems for increments of renewal processes, *Zap. Nauchn. Semin. POMI* **298**, pp. 208–225, english transl.: J. Math. Sci., **128**, no. 1, 2614–2624, (2005).

Frolov, A. N. (2004a). On the law of the iterated logarithm for increments of sums of independent random variables, *Zap. Nauchn. Semin. POMI* **320**, pp. 174–186, english transl.: J. Math. Sci., **137**, no. 1, 4575–4582, (2006).

Frolov, A. N. (2004b). Strong limit theorems for increments of sums of independent random variables, *Zap. Nauchn. Semin. POMI* **311**, pp. 260–285, english transl.: J. Math. Sci., **133**, no. 3, 1356–1370, (2006).

Frolov, A. N. (2004c). Universal strong laws for increments of processes with independent increments, *Theor. Probab. Appl.* **49**, 3, pp. 601–609, english transl.: Theor. Probab. Appl., **49**, no. 3, 531–540, (2005).

Frolov, A. N. (2005). Converses to the Csörgő–Révész laws, *Statist. Probab. Lett.* **72**, pp. 113–123.

Frolov, A. N. (2007). Limit theorems for increments of compound renewal processes, *Zap. Nauchn. Semin. POMI* **351**, pp. 259–283, english transl.: J. Math. Sci., **152**, no. 6, 944–957, (2008).

Frolov, A. N. (2008). Asymptotic behaviour of probabilities of moderate deviations, *Trudy St.Peterburgskogo Mat. Obschestva* **14**, pp. 197–211, english transl.: Amer. Math. Soc. Translations. Ser. 2, **228**, Proc. St.Petersburg Math. Soc., vol. 14, 157–168, (2009).

Frolov, A. N. (2014). *Limit theorems of probability theory* (Izd. St.Petersburg State Univ., St.Petersburg), (In Russian).

Frolov, A. N. and Martikainen, A. I. (1999). On the length of the longest increasing run in R^d, *Statist. Probab. Lett.* **41**, pp. 153–161.

Frolov, A. N. and Martikainen, A. I. (2001). On the largest cave in a random field, *Studia Sci. Math. Hungar.* **37**, pp. 213–223.

Frolov, A. N., Martikainen, A. I. and Steinebach, J. (1997). Erdős–Rényi–Shepp type laws in non-i.i.d. case, *Studia Sci. Math. Hungar.* **33**, pp. 127–151.

Frolov, A. N., Martikainen, A. I. and Steinebach, J. (1998). On the maximal gain over head runs, *Studia Sci. Math. Hungar.* **34**, pp. 165–181.

Frolov, A. N., Martikainen, A. I. and Steinebach, J. (2000a). On the maximal excursion over increasing runs, *In: Asymptotic methods in probability and statistics with applications. Birkhäuser, Boston*, pp. 225–242.

Frolov, A. N., Martikainen, A. I. and Steinebach, J. (2000b). Strong laws for the maximal gain over increasing runs, *Statist. Probab. Lett.* **50**, pp. 305–312.

Galambos, J. (1978). *The asymptotic theory of extreme order statistics* (John Wiley and sons, New York etc.).

Gut, A. (2009). *Stopped random walk, (2nd edn.)* (Springer Science+Business Media, New York).

Hanson, D. L. and Russo, R. P. (1983). Some results on increments of Wiener process with applications to lag sums of independent identically distributed random variables, *Ann. Probab.* **11**, pp. 609–623.

Hanson, D. L. and Russo, R. P. (1985). Some limit results for lag sums of independent, non-i.i.d. random variables, *Z. Wahrsch. verw. Geb.* **68**, pp. 425–445.

Hartman, P. and Wintner, A. (1941). On the law of the iterated logarithm, *Amer. J. Math.* **63**, pp. 169–176.

Höglund, T. (1979). A unified formulation of the central limit theorem for small and large deviations from the mean, *Z. Wahrsch. verw. Geb.* **49**, pp. 105–117.

Ibragimov, I. A. and Linnik, Y. V. (1971). *Independent and stationary sequences of random variables* (Volters-Noordhoff, Groningen).

Kalinauskaĭte, N. (1971). Upper and lower functions for sums of independent random variables with stable limit distributions, *Select. Transl. Math. Statist. Probab.* **10**, pp. 305–314.

Kesten, H. (1970). The limit points of a normalized random walk, *Ann. Math. Statist.* **41**, pp. 1173–1205.

Khintchine, A. Y. (1929a). Über dei positiven und negativen abweichungen des arithmetischen mittels, *Math. Ann.* **101**, pp. 381–385.

Khintchine, A. Y. (1929b). Über einen neuen grenzwertsatz der wahrscheinlichkeitsrechnung, *Math. Ann.* **101**, pp. 745–752.

Kim, L. V. and Nagaev, A. V. (1975). On the asymmetrical problem of large deviations, *Theor. Probab. Appl.* **28**, pp. 58–68, english transl.: Theor. Probab. Appl., **20**, no. 1, 57–68, (1975).

Klass, M. J. (1976). Toward a universal law of the iterated logarithm. I, *Z. Wahrsch. verw. Geb.* **36**, pp. 165–178.

Klass, M. J. (1977). Toward a universal law of the iterated logarithm. II, *Z. Wahrsch. verw. Geb.* **39**, pp. 151–165.

Kolmogorov, A. N. (1929). Über das gesetz des iterierten logarithmus, *Math. Ann.* **101**, pp. 126–135.

Komlós, J., Major, P. and Tusnády, G. (1975). An approximation of partial sums of independent rv's, and the sample df. I, *Z. Wahrsch. Verw. Geb.* **32**, pp. 111–131.

Komlós, J., Major, P. and Tusnády, G. (1976). An approximation of partial sums of independent rv's, and the sample df. II, *Z. Wahrsch. Verw. Geb.* **34**, pp. 33–58.

Lamperti, J. (1996). *Probability: a survey of mathematical theory, (2nd edn.)* (John Wiley and Sons, New York).

Lanzinger, H. (2000). A law of the single logarithm for moving averages of random variables under exponential moment conditions, *Studia Sci. Math. Hungar.* **36**, pp. 65–91.

Lanzinger, H. and Stadtmüller, U. (2000). Maxima of increments of partial sums for certain subexponential distributions, *Stoch. Processes Appl.* **86**, pp. 307–322.

Lin, Z. Y. (1990). The Erdős–Rényi laws of large numbers for non-identically distributed random variables, *Chin. Ann. Math.* **11B**, pp. 376–383.

Lin, Z. Y., Choi, Y. K. and Hwang, K. S. (2001). Some limit theorems on the increments of a multi-parameter fractional Brownian motion, *Stoch. Anal. Appl.* **19**, pp. 499–517.

Lin, Z. Y., Lu, C. R. and Shao, Q.-M. (1991). Contribution to the limit theorems, *Contemporary Math.* **118**, pp. 221–237.

Linnik, Y. V. (1961a). Limit theorems for sums of independent random variables taking into account large deviations. I, *Theor. Probab. Appl.* **6**, pp. 145–163, english transl.: Theor. Probab. Appl., **6**, no. 2, 131–148, (1961).

Linnik, Y. V. (1961b). Limit theorems for sums of independent random variables taking into account large deviations. II, *Theor. Probab. Appl.* **6**, pp. 377–391, english transl.: Theor. Probab. Appl., **6**, no. 4, 345–360, (1961).

Linnik, Y. V. (1961c). On the probability of large deviations for the sums of independent variables, *Proc. 4th Berkeley Symp. on Math. Statist. and Probab., 2, Univ. Calif. Press, Berkeley and Los Angeles*, pp. 289–306.

Linnik, Y. V. (1962). Limit theorems for sums of independent random variables taking into account large deviations. III, *Theor. Probab. Appl.* **7**, pp. 121–134, english transl.: Theor. Probab. Appl., **7**, no. 2, 115–129, (1962).

Lipschutz, M. (1956a). On strong bounds for sums of independent random variables which tend to a stable distribution, *Trans. Amer. Math. Soc.* **81**, pp. 135–154.

Lipschutz, M. (1956b). On the magnitude of the error in the approach to stable distributions, *Indag. Math.* **18**, pp. 281–294.

Loève, M. (1963). *Probability theory, (3rd edn.)* (Van Nostrand, Princeton).

Lynch, J. (1983). Some comments on the Erdős–Rényi law and a theorem of Shepp, *Ann. Probab.* **11**, pp. 801–802.

Martikainen, A. I. (1980). Converse of the law of the iterated logarithm for random walk, *Theor. Probab. Appl.* **25**, 2, pp. 364–366, english transl.: Theor. Probab. Appl., **25**, no. 2, 361–362, (1981).

Martikainen, A. I. (1985). On one-sided law of the iterated logarithm, *Theor. Probab. Appl.* **30**, 4, pp. 694–705, english transl.: Theor. Probab. Appl., **30**, no. 4, 736–749, (1986).

Mason, D. M. (1989). An extended version of the Erdős–Rényi strong law of large numbers, *Ann. Probab.* **17**, pp. 257–265.

Michel, R. (1976). Nonuniform central limit bounds with applications to probabilities of deviations, *Ann. Probab.* **4**, 1, pp. 102–106.

Mijnheer, J. L. (1972). A law of the iterated logarithm for the asymmetric stable law with characteristic exponent one, *Ann. Math. Statist.* **43**, pp. 358–362.

Mijnheer, J. L. (1974). *Sample path properties of stable processes* (Mathematish Centrum, Amsterdam).

Millar, P. W. (1972). Some remarks on asymmetric processes, *Ann. Math. Statist.* **43**, pp. 597–601.

Nagaev, A. V. (1969). Limit theorems including the effect of large deviations when Cramér's condition is violated, *Izv. AN UzSSR, seriya fiz.-matem. nauk* , 6, pp. 17–22, (In Russian).

Nagaev, A. V. (1983). On asymmetric problem of large deviations when the limit law is stable, *Theor. Probab. Appl.* **28**, pp. 637–645, english transl.: Theor. Probab. Appl., **28**, no. 4, 670–680, (1984).

Nagaev, S. V. (1965). Some limit theorems for large deviations, *Theor. Probab. Appl.* **10**, pp. 231–254, english transl.: Theor. Probab. Appl., **10**, no. 2, 214–235, (1965).

Nagaev, S. V. (1981). On the asymptotic behaviour of one-sided large deviation probabilities, *Theor. Probab. Appl.* **26**, pp. 369–372, english transl.: Theor. Probab. Appl., **26**, no. 2, 362–366, (1982).

Novak, S. Y. (1992). Longest runs in a sequence of m-dependent random variables, *Probab. Theory Relat. Fields* **91**, pp. 269–281.

Ortega, J. and Wschebor, M. (1984). On the increments of the Wiener process, *Z. Wahrsch. Verw. Geb.* **65**, pp. 329–339.

Osipov, L. V. (1972). On probabilities of large deviations of sums of independent random variables, *Theor. Probab. Appl.* **17**, 2, pp. 320–341, english transl.: Theor. Probab. Appl., **17**, no. 2, 309–331, (1973).

Petrov, V. V. (1954). Generalization of Cramér's limit theorem, *Uspekhi mat. nauk* **9**, 4, pp. 195–202, english transl.: Selected Transl. Math. Statist. Probab., **6**, 1–8, (1966), AMS, Providence, RI.

Petrov, V. V. (1961). On large deviations of sums of random variables, *Vestnik Leningrad Univ.* , 1, pp. 25–37, (In Russian).

Petrov, V. V. (1963). Limit theorems for large deviations under violation of Cramér's condition. I, *Vestnik Leningrad Univ.* , 19, pp. 49–68, english transl. in "Selected Transl. on Math. and Probab." **7**, 235–280, (1968).

Petrov, V. V. (1964). Limit theorems for large deviations under violation of Cramér's condition. II, *Vestnik Leningrad Univ.* , 1, pp. 58–75, english transl. in "Selected Transl. on Math. and Probab." **7**, 235–280, (1968).

Petrov, V. V. (1965). On probabilities of large deviations of sums of independent random variables, *Theor. Probab. Appl.* **10**, 2, pp. 310–322, english transl.: Theor. Probab. Appl., **10**, no. 2, 287–298, (1965).

Petrov, V. V. (1968). Asymptotic behaviour of probabilities of large deviations, *Theor. Probab. Appl.* **13**, 3, pp. 432–444, english transl.: Theor. Probab. Appl., **13**, no. 3, 408–420, (1968).

Petrov, V. V. (1975). *Sums of independent random variables* (Springer, New York etc.).

Petrov, V. V. (1987). *Limit theorems for sums of independent random variables* (Nauka, Moscow), (In Russian).

Petrov, V. V. (1995). *Limit theorems of probability theory* (Oxford University Press, New York).

Pfuhl, W. and Steinebach, J. (1988). On precise asymtotics for the Erdős–Rényi increments of random fields, *Pub. Inst. Stat. Univ. Paris* **33**, pp. 49–66.

Pittel, B. G. (1981). Limiting behaviour of a process of runs, *Ann. Probab.* **9**, pp. 119–129.

Pruitt, E. W. (1981). General one-sided laws of the iterated logarithm, *Ann. Probab.* **9**, pp. 1–48.

Révész, P. (1983). Three problems on the lengths of increasing runs, *Stoch. Process. Appl.* **15**, pp. 169–179.

Révész, P. (1990). *Random walk in random and non-random environment* (World Scientific Publ., Singapore).

Rosalsky, A. (1980). On the converse of the iterated logarithm law, *Sankhya* **A 42., N. 1–2**, pp. 103–108.

Rozovskii, L. V. (1981). Limit theorems for large deviations in a narrow zone, *Theor. Probab. Appl.* **26**, pp. 834–845, english transl.: Theor. Probab. Appl., **26**, no. 4, 834–845, (1982).

Rozovskii, L. V. (1989a). An estimate for probabilities of large deviations, *Matem. Zametki* **42**, pp. 145–156, english transl.: Mathematical Notes, **42**, no. 1, 590–597, (1987).

Rozovskii, L. V. (1989b). Probabilities of large deviations of sums of independent random variables with common distribution function in the domain of attraction of the normal law, *Theor. Probab. Appl.* **38**, pp. 686–705, english transl.: Theor. Probab. Appl., **34**, no. 4, 625–644, (1989).

Rozovskii, L. V. (1993). Probabilities of large deviations on the whole axis, *Theor. Probab. Appl.* **38**, pp. 79–109, english transl.: Theor. Probab. Appl., **38**, no. 1, 53–79, (1993).

Rozovskii, L. V. (1997). Probabilities of large deviations for sums of independent random variables with a common distribution function from the domain of attraction of an asymmetric stable law, *Theor. Probab. Appl.* **42**, pp. 496–530, english transl.: Theor. Probab. Appl., **42**, no. 3, 454–482, (1998).

Rozovskii, L. V. (1999). A lower bound for probabilities of large deviations of sums of independent random variables with finite variations, *Zap. Nauchn. Semin. POMI* **260**, pp. 218–239, english transl.: J. Math. Sci., **109**, no. 6, 2192–2209, (2002).

Rozovskii, L. V. (2001). On lower bound for probabilities of large deviations of sample mean under Cramér's condition, *Zap. Nauchn. Semin. POMI* **278**, pp. 208–224, english transl.: J. Math. Sci., **118**, no. 6, 5624–5634, (2003).

Rozovskii, L. V. (2003). Probabilities of large deviations for some classes of distributions satisfying Cramér's condition, *Zap. Nauchn. Semin. POMI* **298**, pp. 161–185, english transl.: J. Math. Sci., **128**, no. 1, 2585–2600, (2005).

Rozovskii, L. V. (2004). Sums of independent random variables with finite variances – moderate deviations and bounds in CLT, *Zap. Nauchn. Semin. POMI* **311**, pp. 242–259, english transl.: J. Math. Sci., **133**, no. 3, 1345–1355, (2006).

Rychlik, Z. (1983). Nonuniform central limit bounds with applications to probabilities of deviations, *Theor. Probab. Appl.* **28**, 4, pp. 646–652.

Samorodnitsky, G. and Taqqu, M. (1994). *Stable non-Gaussian processes* (Chapman and Hall/CRC, New York).

Scherbakova, O. E. (2003). Rate of convergence of increments for random fields,

Zap. Nauchn. Semin. POMI **298**, pp. 304–315, english transl.: J. Math. Sci., **128**, no. 1, 2669–2676, (2005).

Scherbakova, O. E. (2004). Convergence rate for large increments of random fields, *Zap. Nauchn. Semin. POMI* **320**, pp. 187–225, english transl.: J. Math. Sci., **137**, no. 1, 4583–4608, (2006).

Seneta, E. (1976). Regularly varying functions, *Lecture Notes in Mathematics* **508**.

Shao, Q.-M. (1989). On a problem of Csörgő and Révész, *Ann. Probab.* **17**, pp. 809–812.

Shao, Q.-M. (1995). On a conjecture of Révész, *Proc. AMS* **123**, pp. 575–582.

Shepp, L. A. (1964). A limit law concerning moving averages, *Ann. Math. Statist.* **35**, pp. 424–428.

Slastnikov, A. D. (1978). Limit theorems for moderate deviation probabilities, *Theor. Probab. Appl.* **23**, pp. 322–340, english transl.: Theor. Probab. Appl., **23**, no. 2, 322–340, (1979).

Slastnikov, A. D. (1984). Narrow zones of normal convergence for sums of non-identically distributed random variables, *Theor. Probab. Appl.* **29**, pp. 570–574, english transl.: Theor. Probab. Appl., **29**, no. 3, 570–574, (1985).

Statulevičius, V. A. (1966). On large deviations, *Z. Wahrsch. verw. Geb.* **6**, pp. 133–144.

Steinebach, J. (1978). On a necessary condition for the Erdős–Rényi law of large numbers, *Proc. Amer. Math. Soc.* **68**, pp. 97–100.

Steinebach, J. (1981). On general versions of Erdős–Rényi laws, *Z. Wahrsch. Verw. Geb.* **56**, pp. 549–554.

Steinebach, J. (1982). Between invariance principles and Erdős–Rényi laws, *Coll. Math. Soc. J. Bolyai* **36**, pp. 981–1005.

Steinebach, J. (1983). On the increments of partial sum processes with multidimensional indices, *Z. Wahrsch. verw. Geb.* **63**, pp. 59–70.

Steinebach, J. (1986). Improved Erdős–Rényi and strong approximation laws for increments of renewal processes, *Ann. Probab.* **14**, pp. 547–559.

Steinebach, J. (1991). Strong laws for small increments of renewal processes, *Ann. Probab.* **19**, pp. 1768–1776.

Steinebach, J. (1998). On a conjecture of Révész and its analogue for renewal processes, *In: Szyszkowicz B. (Ed.), Asymptotic methods in probability and statistics. ICAMPS 197. Amsterdam, North Holland/Elsevier*, pp. 311–322.

Strassen, V. (1964). An invariance principle for the law of the iterated logarithm, *Z. Wahrsch. Verw. Geb.* **3**, pp. 211–226.

Terterov, M. N. (2011). On limiting behaviour of increments of sums of independent random variables from domains of attractions of asymmetric stable laws, *Vestnik St.Petersburg Univ., Ser. 1* , 2, pp. 95–103, english transl.: Vestnik St.Petersburg Univ. Math., **44**, no. 2, 147–154, (2011).

Wolf, V. (1968). Some limit theorems for large deviations, *Doklady AN SSSR* **178**, 1, pp. 21–23, english transl.: Soviet. Math., Doklady **9**, no. 1, 14–17, (1968).

Wolf, V. (1970). Some limit theorems for large deviations of sums of independent random variables, *Doklady AN SSSR* **191**, 6, pp. 1209–1211, english transl.:

Soviet. Math., Doklady **11**, no. 2, 509–512, (1970).

Zinchenko, N. M. (1987). Asymptotic for increments of stable stochastic processes with jumps of one sign, *Theor. Probab. Appl.* **32**, 4, pp. 793–796, english transl.: Theor. Probab. Appl., **32**, no. 4, 724–727, (1987).

Zinchenko, N. M. (1994). On the asymptotic behaviour of increments of certain classes of random fields, *Theor. Probab. Math. Statst.* **48**, pp. 7–11.

Zolotarev, V. M. (1962). On a new point of view on limit theorems taking into account large deviations, *Trudy VI Vsesouyuzn. soveschaniya po teorii veroatn. i matem. statist. Gos. izd. polit. i nauchn. liter. LitSSR, Vilnius*, pp. 43–47, (In Russian).

Author Index

General Index

Printed in the United States
By Bookmasters